GREEN
FIRE

GREEN FIRE

The Life Force, from the Atom to the Mind

Ignacio Martínez
Juan Luis Arsuaga

Translated by Michael B. Miller

THUNDER'S MOUTH PRESS • NEW YORK

Green Fire
The Life Force, from the Atom to the Mind
Ignacio Martínez and Juan Luis Arsuaga
Translated by Michael B. Miller

Originally published as *Amalur: Del átomo a la mente*

Published in the United States by:
Thunder's Mouth Press
245 W. 17th St., 11th floor
New York, NY 10011

First printing September 2004

Library of Congress Cataloging-in-Publication Data:

Martínez, Ignacio.
Arsuaga, Juan Luis.
[Amalur: Del átomo a la mente. English]
Green Fire: The Life Force, from the Atom to the Mind / by Ignacio Martínez and Juan Luis Arsuaga; translated by Michael B. Miller
p.cm
Includes bibliographical references and index.
ISBN 1-56858-307-9 (cloth)

10 9 8 7 6 5 4 3 2 1

Printed in Canada
Typeset and design by Paul Paddock
Distributed by Publishers Group West

CONTENTS

To Our

Professors

and

Students

"The Earth, and properly so, did receive the name
of Mother, for all living beings were nurtured
by Earth itself."

Titus Lucretius Carus, *On The Nature of Things*

"The Earth possesses the vital force which is the basis of the plant kingdom, which invigorates the human organism and is the arbitrator of certain formulas or magical signs, and is that which assures the preservation of the herd if some of the cattle should be offered up or sacrificed to it."

José Miguel de Barandiarán, *Basque Mythology*

Acknowledgments

In expanding on this book, we had the good fortune of receiving suggestions and opinions from a good number of colleagues, specialists in different fields who have enriched the text and corrected inaccuracies. The chapter "A Little Bit of Physics and Chemistry" was revised by Rosa González and Jesus Pérez-Gil, who also made corrections to the chapter "The Devourers of Light." Carmen Roldán, Patricio Domínguez, and Diego García-Bellido revised the chapter "The First of 'Us'" and Manuel Martín Loeches did the same for "Queen Christina's Clock." The chapter "The Origin of Life" greatly benefited from the valuable suggestions and contributions of Juli G. Peretó and Federico Morán.

Our friends Gloria Cuenca, Ana Gracia, Carlos Lorenzo, and Nuria García provided a critical reading of the original manuscript, and their opinions have substantially improved the contents of this book. Moreover, Carlos Lorenzo collaborated on some of the illustrations.

To María Victoria Romero and Pedro María Arsuaga, we owe a great deal for the time they spent in providing us their indispensable documentary support.

A special mention is owed the director of the Biblioteca Histórica de la Universidad Complutense de Madrid, Marqués de Valdecilla, for his timely help in locating many of the illustrations that accompany this text.

Finally, we also wish to express our gratitude to Enrique Bernárdez and Ernest Yellowhair Toppah, director of the Kiowa Tribal Museum, for having recovered some words that had been lost to time.

Introduction

The Colors of the Desert

This is a book about the origin of life and was conceived in that very place where life displays all its grandeur: a desert. The word desert evokes in us the yellowish color of the sand dunes. In fact, this is how the Dictionary of the Royal Academy (*Diccionario de la Real Academia Española*) defines the term desert: "a sandy or stony terrain that, because of the almost total lack of precipitation, is devoid of vegetation or has very little of it." But from the point of view of a biologist, the mere absence or scarcity of vegetation is what really defines a desert. In those places where plants do not grow, animals do not flourish. For a biologist, that is the essence of a desert: the absence or scarcity of life.

From this perspective, there are more deserts than just those with *yellow* sand. It isn't only the lack of rainfall, or the sandy or stony nature of a terrain, which can limit the growth of plant life. *Blue* is the color of the more extensive desert of our planet. It is the open sea. Despite the platitude that the sea is a rich garden, the source of humanity's future provisions, the very color of the seas betrays their real nature. *Green* is the color of life, not blue.

Survival for the vast majority of our planet's living organisms depends on the ability of other specific organisms (basically plants and a heterogeneous mixture of microorganisms that includes single-cell algae, bacteria, and cyanobacteria) to convert water and carbon dioxide into organic matter. To accomplish that, they use the energy of sunlight in a process known as *photosynthesis*, which is the key element found in a molecule called *chlorophyll* (in fact, there are several types of chlorophyll), which is green in color. If the oceans of our planet are bubbling with life, they ought to be green, the color of photosynthesized organisms, of chlorophyll.

The sea's blue color is that of the absence of life, a different color from that of the desert. It is a desert that is neither stony nor sandy, nor a place where vegetation is absent because of any scarcity of rain. Apparently the sea is an ideal place for the growth of plant life: there is an unlimited quantity of water, carbon dioxide, and sunlight. What then impedes the growth of plant life in the ocean? The answer lies in mineral salts. Every farmer, gardener, and garden shop owner knows that for plants to thrive, water and light are not enough. If the earth and oceans are to be fertile, they have to contain, among other mineral salts, nitrogen, phosphorus, and sulfur. The use of fertilizers is justified, precisely because of the need to replenish farmlands and gardens with these chemical nutrients. Without them, plant life cannot synthesize most of the substances they need in order to thrive.

The life of photosynthesized organisms is confined to a thin layer on the oceanic surface. Sunlight is quickly extinguished when it pierces the water and, at a depth of about 100 meters where photosynthesis is not possible, darkness takes over. Thousands of tons of mineral salts—dragged from the continental shelves by the rivers—reach the oceans every year. A portion ends up dissolved in the oceanic waters, conferring upon them their particular salty taste, but what is clear is that the majority of these mineral salts are deposited onto the muddy ocean depths, thousands of meters from the sunlit surface.

Given the quantity of nutrients left to dissolve in the water, only a few organisms can survive in the sunlit zone, forming a tenuous and disperse green film that cannot hide the ocean's blue color or sustain animal life. Of course, there are places where the sea is green, where photosynthesizing organisms provide the basis for animal life to flourish. In some regions, ascendant underwater currents exist, thus pulling the nutrients, trapped on the bottom, to the surface. In other regions, it is the rivers that flood their outlets with mineral salts. In all these places, the sea is green and bubbling with animal life, but these oases represent only a small fraction of the total oceanic surface. It is precisely in these locations where man's fishing activities are concentrated. Outside of these fertile spots, in most of the sea, fishing is not possible for the simple reason that there are no fish. In other words, they form the planet's largest desert: the blue desert.

There are other regions in our world that are not covered by the green blanket of life either; these are the polar caps and summits of tall mountains: our *white* deserts. Here, the factor that impedes plant life is the extreme cold. Plants cannot thrive where water is frozen. Under these conditions, plant juices cannot circulate and, as if this were not enough, the mineral floor is beyond their reach, covered with a thick layer of impenetrable ice, which can extend several kilometers in thickness in the far reaches of Greenland or Antarctica.

Yellow, blue, and white are the colors of our planet's deserts that can be distinguished from outer space. But a fourth type of desert exists that is not visible from the air: the oceanic floors and interior of caves, our *black* deserts. The absence of light is their common denominator, and the reason for their barrenness:

without light there is no photosynthesis. Plants cannot grow in the absolute and endless darkness of caves or oceanic abysses. One might think that the absence of light is the most fatal of life's obstacles, but this is not so. In certain regions of the ocean depths, there are prosperous ecosystems composed of bacteria and invertebrates that do not derive their energy source from solar light, but rather from a species of chemical *manna*, which arises from small underwater volcanoes called fumaroles. From these fiery mouths gushes a stream of boiling water, rich in mineral composites that certain bacteria use as an energy source. These, in turn, are pasture for the invertebrates that inhabit areas around the fumaroles.

But in caverns, there are no fumaroles to replace the vital role of sunlight. Can life also thrive in these locations? As incredible as it may seem, the answer is yes. Every summer, when we unearth buried human remains from over 350,000 years ago, in the Sima de los Huesos dig of the Cueva Mayor, at the Sierra de Atapuerca site, we sometimes come across a few tiny color-blind arthropods that live in the darkness of the cave. They feed off any least particle of organic matter that, by whatever means, filters down to the depths of their dwelling. If we should leave behind one small fragment of wood in the cave at the conclusion of an excavation project, upon our return the following year, we find it covered in a white shroud: fungi that flourish on abandoned wood.

Seeing these tenacious inhabitants of the dark has always left a deep impression on us; they are a testament to the extraordinary ability of living organisms to adapt to and fill in any fissure or nook and cranny of our planet. In Steven Spielberg's famous film *Jurassic Park*, a duel of personalities is established between a paleontologist, Dr. Alan Grant, and a mathematician, Dr. Ian Malcolm. No matter how much our sympathies may lie with the paleontologist, the mathematician has a line in the film that sums up all of our feelings when we contemplate those inhabitants of the dark: "Life finds a way."

This fascination with the phenomenon of life is what guides our research. Our interest goes beyond the knowledge of our more or less direct ancestors. As paleontologists, what interests us is the very phenomenon of life: its origin, evolution, and diversification.

The Three Marvels

Besides having the good fortune to be able to approach the study of life from our investigations in the field of human evolution, we have another privilege, perhaps even greater. Both of us are professors. For many years, we have had the opportunity to exchange ideas and experiences on the teaching of biology and paleontology, from secondary education through university studies. Our students have lent us, and lend us now, ever-fresh eyes for old problems.

We are so accustomed to living surrounded by the marvels of our technological civilization that our ability to be astonished has been numbed. However, to the eyes of a four-year-old, our world is replete with extraordinary and incredible things (such as escalators, remote control devices, or radio-controlled cars). If we remain unmoved before the *miracles* of technology, even more unnoticed are the wonders of the natural world that pass before us. However, every day of class, we see proof of a triple miracle.

In the first place, we never cease to be amazed by the astonishing fact that laws govern nature. This, which seems so obvious, probably has no reason to be so. Our universe could be chaotic, and its phenomena could react to different stimuli at every opportunity. Although, if that were its nature, we would not be here to realize it. But even more surprising is the fact that many of these laws are extraordinarily simple. For example, the reciprocal attraction that bodies experience owing to their masses (the familiar behavior of gravity), which determines events as disparate as tides or the movements and trajectories of stars, is described by means of a simple mathematical equation that offers no complication whatever. Think for a moment that things could be different in a universe different from ours, and the orbit of every planet could react to causes different from those that determine other celestial bodies. Or imagine that gravity was such an intricate phenomenon that it would require very complex equations for us to describe it. If it were capable of discerning those alternatives, it would share with us the wonder of its existence and the simplicity of the laws of nature.

The second astonishing circumstance that normally goes unnoticed is the extraordinary ability of the human brain to know and understand nature. From the infinitely small to the unimaginably grand, we as humans have been able to penetrate the darkest secrets of our universe. We are able to know, with a reasonable level of certainty, events that took place billions of years ago, or that take place hundreds of millions of miles away. We have *tamed* fire, electricity, and even atomic energy to a certain degree. Our knowledge of some of the fundamental secrets of life is so profound that we feel dizziness and restlessness with our ability to manipulate them. Does this not surprise you? People, not machines or powerful computers, have carried out all those discoveries. Progress in scientific knowledge is, perhaps, the human being's most thrilling and dazzling intellectual adventure (and, at times, also just that, an adventure).

As if this were not enough, the practice of teaching constitutes another *miracle*, the third marvel. If it surprises us that nature is constrained by laws, if we are impressed by the fact that we are capable of comprehending them, what should we say about the fact that we can transmit them to one another, without any more help (in the majority of cases) than a blackboard and chalk? The human mind is endowed with an extraordinary instrument, language, which enables us to communicate to each other all kinds of information. This instrument not only includes the different languages, but also the language of nature, which we have been equally capable of learning: mathematics. Together with this priceless instrument, we humans have an enormous capacity for learning, which allows us to comprehend and assimilate those things that our predecessors discovered before us.

Our aim in this text is to take these three marvels beyond the classroom. To do that, we have selected those aspects of our knowledge of life that seem the most fascinating to us. Without a doubt, numerous other topics, equally exciting, have been left outside the pages of this book. But we preferred to plumb the depths on just a few of them, rather than superficially extend ourselves to many areas. We have also included, in all the chapters, a brief history of some discoveries, along with our view on certain aspects of the personalities of the leading characters. We have exercised the greatest care in explaining all questions in the clearest possible manner, while attempting to avoid oversimplification. Perhaps, in the end, this may not be an easy book after all, but we hope it is indeed intelligible.

In the first chapter, we have compiled a brief history of the development of modern science, from the baroque period to present-day. In the second chapter, the history of ideas is further expanded upon and has a valid protagonist. As you might expect, the subject is Darwin, the father of modern biology. In our judgement, the vital contribution that Darwin made to our higher knowledge of living matter makes him worthy of this distinction. We have devoted chapters four to seven to those aspects of living organisms that attract our attention the most: the origin of life, the process of photosynthesis, the relationship between fungi and other organisms, and the appearance and diversification of animals, including the first invertebrates. We have not forgotten chapter three, but this section, which relates to questions in the fields of physics and chemistry, deserves a commentary all its own.

In order to be able to understand the subtlety of the fundamental biological processes of life, it is indispensable for us to know some of the physical and chemical properties of the matter that comprises living organisms. Perhaps we should include the proviso that if you are one who is not fond of physics and chemistry, you should forego this chapter. Nevertheless, we recommend the opposite. The phenomenon of life cannot be understood and appreciated without its chemical-physical basis. The very existence of living organisms involves a challenge (obviously) to some of physic's basic laws, and many biological processes embody astonishing solutions to complex chemical problems. But, be that as it may, knowledge of the physical and chemical properties of matter is a thrilling experience.

In the remaining chapters, from seven on, we have concerned ourselves with the nature and origin of something very difficult to define, but which we all have in common as human beings: the *mind*. Ours is a biological perspective on the problem, and because of it, we begin with the description of the central nervous system and analysis of animal behavior in order to take up the very thorny issue of the biological foundations of our own behavior.

At the end of the book, please look at the bibliographical section. A substantial part of the material in this book has been gathered from other sources, to which we are indebted. There, you will find the compilation of our indebtedness. If one of the topics has piqued a special interest for you, you will also find splendid readings for additional research.

Finally, do not search for many human fossils on the pages of this book. This one time, we have retired those *old* friends of ours from star billing.

CHAPTER I

GOD'S LETTER TO HUMANITY

THE ORDER OF THE WORLD

The return path out of Sima de los Huesos (literally the "Bone Pit") culminates in a steep slope formed by a cone of limestone slabs. At the incline's highest point, the floor of the cave rises and almost touches the ceiling. If the same thing had happened in the Cueva Mayor and all its tunnels, they would not have been discovered.

It often happens that the buildup of deposits hides the entrances to caves and renders them inaccessible to the rest of the network of channels, known as the karst system. The sites that are being excavated at Atapuerca's Trinchera del Ferrocarril are of that type: mouths of caves closed by many tons of sediment. One of them is the famous Gran Dolina, with human fossils 800,000 years old, the first inhabitants known in Europe; another of the caves (called Sima del Elefante), which has older animal fossils (more than a million years old), was one of the entrances, now closed, to the network of tunnels in the Cueva Mayor.

Once past the bottleneck, it opens into a large room, known as the Portalón. The Portalón, in turn, is located at the foot of a great fissure, or crack, in the side of the mountain, which runs almost unnoticed to the top. That's why it is an ideal place to take refuge, cool when the heat outside is unbearable, and warm during the coldest months. There is proof that the people of the Neolithic Period and Bronze Age resided here, and it is also possible that they occupied it in the Paleolithic Age. The excavations, which are underway, will one day tell us. From the Portalón, another long tunnel, called the Galería del Sílex, branches off and was used for funereal purposes by the residents of the Neolithic Period and Bronze Age. In order to exit through the lateral opening, you ascend a short but steep path squeezed between two walls and hidden from view until you reach the summit. This trail up the side of Sima de los Huesos is long and fatiguing (altogether, more than 197 feet), but it affords a view of a dazzling landscape after so much darkness.

The limestone skirts, covered by a forest of holm and gall oaks, of the Sierra de Atapuerca are on the first level. During the spring and summer months, the gall oaks are barely distinguishable from the former; but during the fall and winter months, the leaves of the galls, unlike those of the holms, wither, yet are not totally lost, and their contour is unmistakable. Below, one can see the Arlanzón River's fertile but limited plain, marked by a basin. Here, there are numerous orchards. The course of the river is visible through the stand of poplar and ash trees that run along it. Spread out between the river and limestone are fields of grain: whitish barley and golden wheat. The soil is soft and covered with deposits of terraced pebbles. Here and there, one sees patches of chestnut oaks (or turkey oaks) that once occupied the entire terrain of what is today cultivated land.

Looking out toward the river, to the south, one can see the cars that travel the highway, crossing the town of Ibeas de Juarros, and far in the distance, two large hills, at whose feet lies the town of

Covarrubias, are visible. To the west lies the great Castilian plateau, and stretched out along the Arlanzón River is the city of Burgos; on a clear day, the spires of its cathedral are distinctly visible. But to the east, the landscape is dominated by the high summits of the Sierra de la Demanda, with the San Millán peak standing out above all the rest. On the last glaciation, the Sierra de la Demanda gave shelter to small glaciers in its deep gorges.

The landscape upon which the Neolithic and Bronze Age farmers and cattle breeders gazed could not have been much different from what we see today, although the geography of chestnut oaks no doubt extended much farther; we ourselves have seen it recede in recent years. The residents of the Neolithic Period were, nevertheless, the first to open clearings in the forest in order to tend cattle and cultivate grain. The ax, fire, and teeth of domestic animals were their allies. Great urban concentrations were nonexistent then, as was, of course, the automobile.

However, the inhabitants of the middle Pleistocene Age, who appeared 350,000 years ago outside the Cueva Mayor, saw the river run much closer; in fact, for thousands of years, the river had been forging its basin in the soft terrain and moving away from the hard limestone of the Sierra. During the various phases of carving itself out, the Arlanzón deposited stones on the plateaus that have been converted into cultivated fields today. On the other hand, the impact of man on the ecosystems was incomparably smaller during the Paleolithic Period.

In the soft twilight of a day at the end of July, a gentle wind stirs the grain in the cornfields, and the landscape produces a pleasant feeling of harmony and placidness.

This impression of standing before a painting, at once perfect and complete, is one expressed by many poets in verses that sing of nature's serene beauty. Everything seems to be in its proper place, fulfilling its destiny. We are overcome with the peace that provides order.

That idea, the one that nature is ordered, has been used as a proof of the existence of a great architect of the cosmos. The very word *cosmos*, from the Greek, means "order." The study of the cosmos, conceived as an ordered whole, whose function and nature can be understood through reason, was what gave rise to Greek philosophy; to discover what was the origin—*arjé*—of the world was the first question posed by the philosophers.

Nature's supple beauty leaves a strong impression on sensitive men, but it is still, quite clearly, a subjective emotion, which only exists so long as there is someone with the sensitivity to perceive it. Beauty requires a spectator in order for it to exist, and it seems that our species has developed a sense that is absent in the rest of the animal kingdom: the sense of "good taste," which provides us our aesthetic pleasures.

On the other hand, order is an objective reality, independent of the spectator, and so it leaves an impression on the minds of enterprising thinkers, philosophers, and lovers of truth. Love of truth, next to love of beauty, are the two qualities that, above all others, set the human being apart.

That's why Saint Thomas Aquinas (1225–1274), in his most famous work, *The Summa Theologica*, presented the order of the world as the fifth *rational* proof of God's existence. It is the proof of the design (or of *theology*): since nature constitutes a harmonious system, in which each element directs itself toward the fulfillment of its function, it must, out of necessity, have an architect. Saint Thomas lived in the thirteenth century, but in religious, scholastic journals, the order of nature continually appears as an indisputable argument in favor of the necessity of a Grand Creator.

But where is the order in the landscape that one gazes upon

from the mouth of the Cueva Mayor, and what is its design? The valley through which the Arlanzón River runs has benefited from the waters that have been flowing through it for hundreds of thousands of years. For that matter, the water has followed two courses: the terrain's general slope and the substratum's hardness. All the waters of the Duero basin flow to the Atlantic, because of late, the plateau has sloped toward the west. Prior to that, it was a closed basin, completely surrounded by mountains, where the sediments that had gradually been accumulating were carried down in runoffs from those mountains. The river has cut off those sediments because they are softer than the limestone of the Sierra de Atapuerca, and therefore, the Arlanzón has been shifting away from it. The stones that it transports and deposits next to the Sierra de Atapuerca come from a great distance, from another mountain range, the Demanda, because the soft sediments do not supply rocks that can be rounded and shaped into pebbles.

In short, the valley was not designed for the river to run through it, but rather it is the result of the laws of physics that determine the course of streams and rivers, in the same way that the burnished rocks of high mountains were not made smooth so that ice could slide down them more easily, but rather they were eroded by glaciers. Nor are rocks rounded so they can be more easily dragged along river bottoms; it was the dragging that shaped them.

Laws, rather than purpose, lie behind the aesthetic imprint of the landscape. Neither do mountain summits have those graceful peaks so they can compete for splendor with the spires of cathedral towers, but rather they are the result of erosion. If it is erosion that shapes the landscape, or rather sculpts it, other forces, internal ones, create the mountainous reliefs we see exhibited upon it. Thus the sense of order disappears, is erased, when we come to understand the mechanisms that have really given shape to what we see.

Underlying all that, of course, is the order of natural laws, which the mathematician can find thrilling, but barely so the poet, since this order appeals more to the head than the heart. Ever since the discoveries of scientists like Copernicus and Galileo, the universe has been presented to us as a system governed by mathematical laws, much colder and more impersonal than a red sunset.

On the Sierra de Atapuerca landscape in 2002, there are a great many man-made inventions. These artificial objects are clearly in response to a plan and have a function to fulfill. The tractor was designed for agricultural work, and the car for transportation. Each one of their parts was conceived with some concrete task in mind and has an assigned duty or use. It performs a specific task: the headlights shine, the motor runs, the wheels turn, the bumpers absorb shocks, etc. The tools that man builds are objects that have a plan and purpose; objects outfitted for a project, as the Nobel Prize–winner Jacques-Lucien Monod (1910–1976) said, or also *teleological* objects (in other words, oriented toward an end), to use a more philosophical word. The stone ax of the prehistoric man of Atapuerca was, of course, an object designed for a task: when we as researchers are faced with the study of a hand-carved stone, we ask ourselves: what was its purpose, and to what end was it constructed?

An essential component of the landscape still remains to be analyzed, and this is the living component. We see plants, and we see animals. Like the machine built by man, they too are composed of structures that accomplish an end. The roots of plants, their vessels, their leaves, serve a nutritional purpose; the flowers are reproductive organs; the seeds will give birth to new plants (they already have, in fact), and they are also surrounded by mechanisms that facilitate their dispersion: some by wind, others in the stomachs or on the skin of animals. Nothing is left to chance.

The eaglet that flies low above the plains has an enormous wingspan that enables it to float across the cornfields with a minimum expenditure of energy; its beak, its talons, are those of a bird of prey; all the systems and organs of its body fulfill some function, all having a use, which is to render a determined service. Everything seems calculated down to the millimeter, with extreme meticulousness, whether it is a mammal, bird, insect, or any other kind of animal. When analyzing the structures of an animal, the zoologist also asks: what is their function? The question is completely justified whether speaking of an eye or camera. The structures of living organisms are adaptive, fitted to their function, in the same way tools are adapted to the use for which they were designed. That is a radical difference between the world of biology and that of geology: the earth's relief formations do not fulfill any function; they are not adaptive: a mountain is not designed to fulfill some plan. While gazing down on a valley, it is pointless to ask: what is its purpose?

Colin Pittendrigh proposed that in order to distinguish them from the teleological objects that man produces (such as the photography camera), the instruments of living organisms (such as the eye) should be designated as *teleonomical* mechanisms. This precise correlation, between the objects equipped for a purpose that we assign to them, and the objects equipped for a purpose in the organic world, goes beyond being utilitarian, because in both cases, they are constructed in accordance with plans that existed prior to the objects themselves. In living entities, the information that determines what they will be resides in the genes: each and every cell in their bodies has a complete plan. Rather than being constructed by the human hand, living entities are self-assembled. And, of course, although their constituent elements have a purpose, in the sense that they realize certain functions, individual entities, as such do not: they do not produce any service and are limited to surviving and reproducing, each one in its own environment and in its own way. To do this, they need to be efficient, and so (and not *just for any reason at all*) adaptations have appeared over time in the course of evolution. Domestic animals and cultivated plants present characteristics that we could properly call teleological; sheep's wool, a cow's enormous udders, and the ear of a cereal plant stuffed with wheat grain have been selected precisely for the purpose of serving man.

The Book of Nature

"The world is the letter that God wrote to humanity," reads a sentence attributed to Plato (c. 427–347 B.C.). The idea is not exclusive to the Greek world. In the Bible (Isaiah 34:4), we read, "the heavens shall be rolled together as a scroll," from the hand of God; even Saint Augustus came to assert that the divine *signature* is to be found somewhere in the universe. In a novel titled *Contact,* the famous astrophysicist and lecturer Carl Sagan (1934–1996) places it in the number pi; you know how it goes: billions and billions . . . etc., etc., etc., and it is after many, many etceteras that somewhere in the universe, at a place as yet unreachable, this signature would be found.

But that letter—or that book—has been read in different ways throughout Western history. The seventeenth century's so-called Scientific Revolution, which led to the birth of modern physics and chemistry and, above all, produced the *scientific method* that we still use, consisted precisely of reading *God's Letter*—which was studied in the medieval and Renaissance universities—in a totally different way from that of Aristotle (c. 384–322 B.C.E.), the medieval scholastic theologian Saint Thomas, and other authors.

For Aristotle, nature's movements are produced because things gravitate toward their natural place in the universe, where they lie in repose once they are there. A stone falls or flames ascend because their places are earth and sky, respectively. Water is another heavy element, like earth, but air, on the other hand, like fire, tends to spiral upward. This explanation serves to explain the *natural* movement of things, which of necessity move in a straight line. Man can conquer nature and leave his ascendant footprint on a stone, or parallel to Earth's surface, but these are forced, artificial movements.

We can find traces of Aristotelian thought about the different nature of the elements in a very beautiful sonnet, in our judgement, one of the most profound and moving poems in the Castilian language, by Lope Félix de Vega Carpio (1562–1635). It is called "To a Gentleman, Bearing His Lady for Burial" ("A un caballero, llevando a su dama a enterrar él mismo"), and it goes like this:

> The sky at his shoulder, though its Sun is fireless,
> And in mortal eclipse, the most beautiful
> Stars, pure roses turned to snow,
> And sky, Earth in unequal habit.
>
> Earth, a forced doom,
> And Atlas, unlike it, sent to cold stone slabs
> Which now await his pitiful garments,
> Sisyphus, you are, through another uncertain summit.
>
> I beg you, tell me, if Love can dare to,
> When was my sadness heavier, Fernando,
> When I am fire or have turned to snow?

Fire is weightless, exhaling
Matter at its core its load is light:
Snow is water and its weeping will weigh heavy.

Al hombro el cielo, aunque su Sol sin lumbre,
Y en eclipse mortal las más hermosas
Estrellas, nieve ya las puras rosas,
Y el cielo, Tierra en desigual costumbre.

Tierra, forzosamente pesadumbre,
Y así no Atlante, a las heladas losas
Que esperan ya sus prendas lastimosas,
Sísifo sois por otra incierta cumbre.

Suplícoos me digáis, si Amor se atreve,
¿cuándo pesó con más pesar, Fernando,
o siendo fuego o convertida en nieve?

Mas el fuego no pesa, que exhalando
La materia a su centro es carga leve:
La nieve es agua y pesará llorando.

Here the great poet is playing with the two meanings of *weight*: the physical and sentimental. In the first quartet, he describes the gentleman carrying on his shoulder the lifeless body (the Sun *grown dark*) of his beloved (the *sky*), whose eyes have closed upon themselves (*the beautiful stars in mortal eclipse*), and whose cheeks have lost their color (*the pure roses now like snow*). In other words, the sky has transformed itself into the earth. In the second quartet, the gentleman is told that he is not the giant Atlas, who held up the earth on his shoulders, according to classic mythology, but rather the unfortunate Sisyphus, condemned by Zeus to endlessly push a rock up a mountain. In the third quartet, the lady's weight in life is compared, when she was fire, with that of her corpse, frozen like snow. In the impressive final three lines, Lope alludes to the Aristotelian idea that fire is weightless and ascends to the heavens, while water is an element with an earthbound proclivity. The final verse, where the two meanings of the verb "to weigh" are juxtaposed against each other is simply unforgettable.

"The Phoenix of Geniuses" studied under the Jesuits in Madrid at the Academy of Saint Isidore (established in 1545) and at the University of Alcalá, acquiring, as is obvious, a sound classical education (in 1603, the Academy of Saint Isidore became the Royal Academy of Madrid and, in 1624, was granted university status; a year later, the name was changed to the Royal Academy of Saint Isidore and, finally, ended up being merged with the University of Alcalá and the Royal Academies of Medicine, Surgery, and Pharmacy in the present-day Universidad Complutense de Madrid).

In summation, Aristotelian thought interpreted the movements of things as the actualization of the potentialities of their constituent elements; that is to say, as if elements themselves had aspirations or desires that might be fulfilled. In other words, elements were treated as if they had a will, as if they were living beings with intentions, biology being the model for Aristotelian physics and chemistry. This type of thought is classified as teleological, which, as we have seen, means "oriented toward an end," that being the key to interpreting the behavior of the totality of nature: inanimate objects, as well as animate beings, aspire to achieve a state of plenitude; there is no substantial difference between the explanations that are given for one or the other.

Rebelling against that seemingly quite logical *animistic* vision of physics and chemistry were the seventeenth-century revolutionary scientists who classified themselves as *modern*. In contrast to the Aristotelians and the Scholastics, they only allow for mechanistic explanations and aspire to express the world's functioning in mathematical language. In fact, they believe, as Plato had already said, God's letter is written in mathematical terms. The Italian Galileo Galilei (1564–1642) proclaimed in his work *Saggiatore* (*Sagittarius*) that the new philosophy of nature must be expressed in mathematical language because mathematics is the structure of nature: "Philosophy has been written in the enormous book we have before our eyes—the Universe. But we can only read it if we know its letters and learns its language. It is written in the language of mathematics and its letters are triangles, circles, and other geometric figures; absent these mediums, it is impossible for the human being to understand even a single word."

Another learned person of the time, the German astronomer Johannes Kepler (1571–1630), affirmed the following: "The Supreme Creator, upon creating this mobile world and stipulating the order of the heavens, abided by the five common principles that have been so well known from the time of Pythagoras and Plato right up to our time and [. . .] in accordance with their nature, he adjusted their number, their size, and the reason for their movements." Those normal geometric bodies are the tetrahedron, cube, octahedron, dodecahedron, and icosahedron, which for Plato were the symbols of the five elements of the universe: fire, earth, air, wind, and water. What makes these five bodies ideal forms is the fact that they are the only ones whose surfaces have equal sides and angles, and whose edges rest upon a sphere. Johannes Kepler, as a consequence, tried to link the ideal bodies with the spheres in which the planets supposedly were placed.

A much more recent physicist, Eugene Paul Wigner (Nobel Prize 1963), expressed the power of mathematics this way to explain the natural world: "The enormous utility of mathematics in the natural sciences is something that brushes up against the mysterious, and there is no explanation for it. It is not at all natural for the 'laws of nature' to exist, and much less so for man to be capable of discovering them. Such a miracle, which results in the language of mathematics for the formulation of the laws of physics, is a marvelous gift that we neither understand nor deserve."

Another interesting aspect of baroque *modern* science (still called *natural philosophy*) is that the idea of the classical atomists (Democritus and Lucretius) is revived, that matter is composed of corpuscles, or in other words, small particles or elements invisible to the naked eye. The existence of corpuscles continued to be a working hypothesis, because no one had seen them. But the important thing is that corporeal attributes (color, taste, temperature, and flexibility), and the changes they experience in solid bodies, must be explained as a consequence of the characteristics and movements of the constituent corpuscles, mathematically described, if possible.

These modern scientists imagined the universe to be a great mechanism and found in the clock the ideal metaphor for how the world functions. A clock is a very complex system of pieces that have been integrated to function harmoniously (although each piece has its own movement) without any need to have a will or purpose attributed to them. It was the clock's regularity that made this mechanism an apt metaphor for the predictable movements of celestial bodies. The new philosophy was thus mechanistic, corpuscular, and aspired to be mathematical in nature.

Moreover, in the same way that the intelligent behavior of the clock's mechanism responded to a design produced by an intelligent being, the clockmaker, also deduced from the universal clock's precision the necessity of the existence of a supernatural architect.

In the seventeenth century, they were quite familiar with automated devices: mechanical artifacts that represented animals and humans, without anyone believing such devices to be endowed with brains or desires. As a consequence, the mechanistic concept of nature was not opposed to Christian doctrine, but instead reinforced it. The *animistic* ideas of the Aristotelians concerning nature can lead some to think that all the things of this world had a conscious life, a dangerous idea, while the metaphor of the clock reinforced the orthodox doctrine of the separation of matter, spirit, and the soul's immortality.

What modern scientists were demanding was the right to read *God's Letter.* If the *Book of Nature* was as divinely inspired as the Holy Scriptures, why shouldn't it be legitimate for natural

Figure 1a: *The modern scientist of the baroque scientific revolution went beyond Aristotle's ideal heaven to discover the true nature of the universe, which he tried to explain in mechanical terms. (Illustration of* The Atmosphere, *Météorologie populaire, by Camille Flammarion, Paris, 1888.)*

philosophers to apply themselves to reading it? Especially when natural philosophy made clear the majesty of the author who designed the great universal machine.

Galileo defended the heliocentric theory (the sun at the center), of the Polish scholar Nicolaus Copernicus (1473–1543), expounded upon in the book *Of Orbital Revolution* (published the same year as the author's death). The Copernican theory that Earth and the planets revolved around the sun challenged the senses, which seemed to indicate otherwise, and the geocentric theory of Claudius Ptolemy (c. 100–170), who asserted that the sun, moon, and planets rotated around Earth. Moreover, one also reads in the Holy Scriptures (Joshua 10: 12–13), for example, that Joshua ordered the sun to stop so that there would be no night, and they could continue the Battle of Gibeon until complete victory, proving that it was the sun that moved:

> Then spake Joshua to the Lord in the day when the Lord delivered up the Amorites before the children of Israel, and he said in the sight of Israel:
> "Sun, stand thou still upon Gibeon;
> and thou, Moon, in the valley of Ajalon!"
> And the sun stood still,
> and the moon stayed,
> until the people had avenged themselves upon their enemies.

Nevertheless, the Copernican theory initially had some success, and in Spain, the Polish scholar was taught at the University of Salamanca. In 1561, the statutes of that university established that in the second year of the School of Astrology, Ptolemy, Geber, or Copernicus could be read at the students' discretion ("upon a vote of the auditors"), and the statues of 1594 decreed that Copernicus be read in the fourth quarter of the second year.

It is possible that at Salamanca, the practical side of Copernicus's very precise astronomical tables drew more interest than his heliocentric theory, but there was one student from Salamanca, an Augustinian monk named Diego de Zúñiga (c. 1536–1598), who tried to make Copernicus's ideas compatible with the Bible. In his commentaries on Job (*In Job Commentaria*) of 1584, Zúñiga defended the superiority of Copernicus over Ptolemy upon explaining the course and positions of the planets, and on the other hand he tried to show that the movement of Earth is not incompatible with the Holy Scriptures. In Ecclesiastes, Salomon affirms that "the Earth abideth for ever," but it must not be interpreted as being stationary, because if the entire sentence is read, it becomes clear that it has a different meaning: "One generation passeth away, and another generation cometh, but the Earth abideth for ever," which is simply to say that Earth does not change the way men do, nor that it is at rest.

Unfortunately, the Inquisition held to the letter of the Bible, and in 1616, Copernicus's book was placed on the *Index librorum prohibitorum*, the nefarious index of books banned by the Church, a ban under which many a living person was yet to suffer (since it lasted from 1559 to 1948). Curiously, although Copernicus, neither in life nor his book, had any issues with the Church, he disappeared from the teachings after 1616, even at the University of Salamanca. Zúñiga's book *In Job Commentaria* was also condemned, by none other than the author himself, in a later work (*Philosophia*, 1597), where he declared the movement of Earth to be an impossibility: "Furthermore, heavy objects that are tossed up into the air, even though the action be repeated a thousand times, will fall back perpendicular upon the same spot; but if the Earth should be propelled upward with equal force, it would be dislodged from the place from which it was launched: thus, it is clear, the Earth does not move."

The Church, especially his Jesuit enemies (who were, to make matters worse, preoccupied with scientific themes), warned Galileo to forget about Copernicus's ideas, which he did for a time, until finally in 1632, he published his *Dialogue on the Two Principal Systems of the World, the Ptolemaic and the Copernican*. As we know, the Inquisition condemned Galileo in the famous trial that took place the following year (1633), in which he was forced to rectify his opinion that Earth was a planet that moved around the sun, and Galileo was obligated to remain in seclusion in his house in Arcetri (near Florence). Here is his repudiation of the idea of the movement of Earth: "I, Galileo, son of Vincenzo Galilei, a Florentine, aged seventy years, arraigned personally before this tribunal, and kneeling before you, most Eminent and Reverend Cardinals, Inquisitors general against heretical depravity throughout the whole Christian Republic [. . .] with sincere heart and unfeigned faith, I abjure, curse, and detest the aforesaid errors and heresies, . . . " Tradition says that when he stood, he stamped the floor with his foot and murmured: "*Eppur si muove!*" ("But all the same, it moves!")

The truth was, the Church did not find it problematic for anyone to use Copernicus's mathematical model in astronomy, but it did not accept the claim that the sun, in fact, and not Earth, occupied the center of the universe. Nevertheless, Galileo did not try to deny the fundamental truth of the Bible, only its literalness, and above all, he maintained that the interpretation that natural philosophy made of the *revealed* word of God in the *Book of Nature* was at least as valid as the interpretation that the theologian made in his own way of the revelation contained in the Scriptures. For the natural philosophers of the baroque period, the world was a theophany, a manifestation or *explicatio* (explanation) of God.

In Spain, the Inquisition had a nefarious effect on science during the baroque period. Any deviation from Aristotle's natural philosophy was suspicious, and in particular, the heliocentric theory was not admissible, despite the fact that some authors resorted to the subterfuge of deeming their reality condemned, but not its theoretic possibility. The *modern* scientists, at the end of the Spanish seventeenth century, known as *novatores* (innovators), certainly did not have it easy when it came to breaking with traditional thinking and constructing a new consciousness, but they formed the basis of Spanish science in the Age of Enlightenment.

To the modern scientists of the seventeenth century, Aristotle's doctrine and that of the Scholastics of *substantial forms* also seemed absurd: the idea that objects consist of matter, but that it is the form that makes them what they are, the same as it is with a sculpture, where it is the shape given to it by the sculptor that transforms the lifeless marble into the likeness of a person. For Aristotle, matter does not enter into the picture, because the cause of the properties of things is in their form. The objects of this world that are called horses differ among themselves (there are no two horses exactly alike), but they share a common *substantial form* that, despite their individual or *accidental* differences, has them belong to the same category of thing and possess a series of qualities that are typical of that class of bodies called horses. The *form* is not *matter*, but is intimately and mysteriously connected to it.

For the modern natural philosophers, the Aristotelian doctrine was unintelligible because it could not be understood in mechanical terms nor could it expect to be expressed in mathematical language. Heat, for example, could be better explained by the means of the movements and collisions of the tiny corpuscles that form

Figure 1b: *Cover of Galileo Galilei's work,* The Dialogue, *on the two principal systems in the world, the Ptolemaic and Copernican, published in 1632, for which the Inquisition condemned him the following year.*

matter, rather than resorting to the mysterious idea of *substantial forms*. For the same reason, modern scientists also contended against a concept of nature that had been developed immediately before, during the Renaissance, and which was based on *occult powers*. This theory defended the idea that bodies were drawn to or repelled by each other by forces that acted from a distance in some mysterious way; the influence of the distant celestial bodies on human beings, which gave foundation to astrology, was a consequence of those *occult powers*. For the mechanistic natural philosophers, Aristotle's theory of *substantial form* also held an implication of *occult powers*.

Mechanism, however, as an explanation, had its limitations, as was soon to be seen. The Englishman Sir Isaac Newton (1642–1727) appeared to be the chief exponent of the Scientific Revolution, but he could still not explain the whole of physics in mechanical terms. Newton's law of universal gravity established the attraction of bodies through a force that acted from a distance, and therefore, even though gravity might be formulated in an irreproachably mathematical way, Newton was accused by philosophers, like the German Gottfried Wilhelm Leibniz (1646–1716), of resorting to *occult powers*. Newton found no reasonableness in the idea that gravity is a force that acts from a distance, and he tried, unsuccessfully, to find a way for the force in question to be mechanically transmitted through a medium. However, even though it had no mechanical character, gravity served to explain reality through a mathematical law, and in that sense, he was a *modern* scientist.

Aristotle's Heaven

"There are more things in heaven and earth than those that are dreamt of in your philosophy," Hamlet tells Horatio, and the new optical instruments of the modern natural philosophers allowed them to look out on worlds previously unknown, for they were either very small or very large. But instruments were not the cause of the Scientific Revolution, rather it was the elaboration of a new group of norms for constructing knowledge, which has been called, since that time, the *scientific method*.

At the outset, the method challenges the *argument of authority*, the *magister dixit*, and it is that critical attitude before the established truth that characterizes the modern scientist. The impostors and deceitful defenders of the *paranormal phenomena* and other nonsense attribute to academic science a rigid and absurd obstinacy. Nothing is farther from the mind of a scientist than deeming as good an idea of another scientist without proof. The fact is that those *paranormal phenomena* are, to be sure, not provable.

In the book *The Legend of Jaun de Alzate*, Pío Baroja (1872–1956) provides a wonderful example of a defiant attitude before the old argument of authority. Jaun is an intelligent villager who mistrusts the classic texts and searches for a truth based on facts. In one passage from the book, Jaun and his friends imagine themselves flying from Larraun Mountain to the Mediterranean:

> *Chiqui*: Let's take off for the Mediterranean, staying parallel to the Pyrenees.
>
> *Macrosophos*: Hold it right there! What do you mean, parallel? That's impossible. Strabo says the Pyrenees run from north to south, in a direction parallel to the Rhine.
>
> *Chiqui*: Strabo's wrong.
>
> *Macrosophos*: Strabo can't be wrong.
>
> *Jaun*: Well, he's wrong. Pliny the Elder knew the Pyrenees run east to west, and Ptolemy marked the position of the two farthest points: one, the

Cape of San Sebastián, in the ocean; the other, the Temple of Aphrodite, of Portus Veneris, in the Mediterranean.

Macrosophos: We have to see who carries more weight, if it's Ptolemy or Strabo.

Jaun: No. We have to see who's telling the truth.

When he built his telescope, Galileo looked out on a new reality that challenged established thinking. In the heavens, Aristotle imagined there was an outer sphere that held the stars, and inside it, a series of integrated spheres made up of a transparent substance (crystal, liquid, or air) holding the planets like visible bodies. The sphere of stars and those of the planets revolved of necessity in circular movements, and the planets had to be flawless spheres, because the circle and sphere are the forms that represent the absolute perfection that reigns in the heavens. Everything is immutable and perfect in the heavens beyond the moon; change and imperfection exist only on Earth.

Lope de Vega made an allusion to that geocentric vision of the universe. In his *El peregrino en su patria* (*The Pilgrim in His Own Country*), we read:

May the Sun, a thousand times slipping, tell,
about the golden parallels of the sky
and make his white sister's face grow large,
the diamonds of her pure veils,
that live fixed in their eighth sphere,
shall not equal me even if they kill me with jealousy.

El Sol mil veces discurriendo cuente
Del cielo los dorados paralelos,

Y de su blanca hermana el rostro aumente,
Que los diamantes de sus puros velos,
Que viven fijos en su otava esfera,
No han de igualarme aunque me maten celos.

The sun revolves around Earth and traverses the parallels along which learned scholars have divided the planet. Their *white sister* is the moon, and the stars (*diamonds of her pure veils*) are fixed in the eighth sphere, with all the spheres revolving around Earth.

When Ptolemy, in the second century, gave mathematical form to the heavens to predict the movements of the planets, he followed Aristotle's model (although, curiously, the trajectories of the planets did not behave as if they were in spheres subject to rotation). But Galileo aimed his telescope at the moon and observed the irregularity of the craters, as imperfect in their *circularity* as those of Earth. Moreover, Galileo saw a comet located beyond the moon, instead of closer, where it should have been if the heavens above the moon were immutable and unchanging. But those observations did not succeed in convincing everybody, because there were people who, as Galileo said, "before modifying anything in Aristotle's heavens, would dare to deny what they observe in the heavens of nature."

The really open minds throughout all of history have been capable of doing without the blinders of pre-existing theories, and that alone, the independence of judgement, is the basis for progress in knowledge. The Spanish astronomer Jerónimo Muñoz (c. 1520–1592) observed in the skies the nova of 1572, a phenomenon that he erroneously attributed to the appearance of a comet instead of what it was in reality, a star that suddenly increased in brilliance. But that is the least of it, because the important thing is that Muñoz was able to deduce that this newly appeared celestial body was beyond the moon, and therefore, the heavens were not immutable (*incorruptible*) as Aristotle believed, but rather subject to changes like Earth itself and all it contains. The method he employed was to calculate the parallax of the *comet* and conclude that it was inferior to that of the moon, meaning it could not be a sublunar body, that is to say, located closer to Earth (the parallax is the difference between the positions that a star has in the sky, depending on the point from which it is observed; the greater the parallax, the smaller the distance, and the smaller the parallax, the greater the distance). Forced to choose between the result of his observations and the Aristotelian model—consecrated by tradition, believed by everyone, and accepted as official doctrine—the Spanish astronomer, like the good scientist he was, stuck to what he could see and measure. These are Jerónimo Muñoz's words in his 1573 treatise, *Book of the New Comet, and of the Place Where They Are Made, and How One May See by the Parallaxes How Distant They Are from the Earth and the Prediction of the Same* (*Libro del nuevo Cometa, y del lugar donde se hazen; y como se vera por las Parallaxes cuan lexos están de Tierra y del Pronostico deste*):

"Therefore, God having granted me the favor of having a free talent, being altogether inclined and well prepared to understand any thought, seeing the weakness in the reasoning with which Aristotle wished to prove the heavens eternal, and having observed with instruments the changes existent in the heavens [. . .] I now believe that what he says to be false, the common opinion being that there has never been any change whatsoever in the heavens, thereby making them immutable. And because I know that changes do take place in the heavens, and comets are born there, I have been forced by natural reasons and geometric signs to concede that alterations and fires do exist in the heavens."

In the narrowest sense, there is no codified scientific method that exists as a group of procedures, even though everybody speaks of it.

The Englishman Francis Bacon (1561–1626) understood scientific method as induction, that is to say, the collection of the greatest possible number of facts, out of which would come the general laws that govern nature. But not all scientists have always been in agreement that regularities emerge from particular cases without any reason, and many have preferred to confront the facts with a previous idea in their heads, that is to say, to move from the general to the specific, to explain individual cases through general laws.

Karl Popper (1902–1994) and other twentieth-century philosophers have tossed this problem around many times, but as scientists, we all agree that there has to be some kind of connection between ideas and facts if one wants to construct a true knowledge of the real world. Mathematics alone can forego observations: in the mathematical method, one begins with the acceptance, without discussion, of certain general principles from which a series of conclusions are logically deduced. But the *exact* sciences, as they are called, concern themselves with worlds that could exist, and the sciences known as *experimental* with the world in which fortunately or unfortunately we live. In the former, there is no room for error, but it is the latter that allows us to read Plato's *God's Letter* and Galileo's *Book of Nature*.

It should be added that what Bacon meant by facts were bona fide observations, carried out by qualified people, and adequately planned experiments. One of the characteristics of modern natural philosophy was the rigor in the collection of data, which must not be influenced by preconceived notions. The new natural philosophy set aside moral judgements and individual interests, and aspired to be objective and dispassionate. But it could even put common sense to the test, like the time it asserted that Earth, and not the sun, was the star that revolved, despite the senses telling us differently.

• • •

THE EXPERIMENT

An important case, in which one traditionally resorted to animist thinking for an explanation, was that of the operation of the pumps that are used to extract water from a well. When the pump creates a vacuum, the water rises through the pipe until it reaches a specific height, which had been observed as being set at more than ten meters. In the 1660s, the Englishman Robert Boyle (1627–1691), who wanted to prove to what degree this was true, constructed in London a metal tube 10 meters long and fitted it with a piece of glass so that he could clearly observe the water rising. In this way, he assured himself that no water had been lost through filtration, and the experience was carried out under the requisite conditions. The column of water rose 10.2 meters.

What was the probable cause for the liquid rising through the tube? The answer could be, according to Aristotelian thought, *in the horror of the void felt by nature*, which probably pushed the water upward *in an effort* to fill the empty tube. In other words, the height attained by the water in its ascent is a measure of the degree of horror that the water experiences because of the vacuum. But, in fact, Aristotle denied that any void could even exist, bearing in mind how much nature abhors it.

In 1644, however, a disciple of Galileo, the Italian Evangelista Torricelli (1608–1647), applied a mechanical principle, that of equilibrium, to solve the problem. There is nothing inside the tube so long as there is air outside. If the air is heavy, and there is no air inside the tube, then the *weight* of the air will push the water up through the tube, until the weight of the column of water is equal to the pressure that the air exercises on the liquid. Mercury is fourteen times denser than water, and if Torricelli was right, it was possible to predict that the mercury would rise in the tube up to a fourteenth part of the height reached by water in suction pumps, which is exactly what happened: the column of mercury rose 73 centimeters. Torricelli, on the other hand, had just invented the barometer, an instrument that allows, in centimeters of mercury, the linear measurement of something as abstract as atmospheric pressure. At the time, Torricelli said, "We live at the bottom of an ocean of an element called air that, through indisputable evidence, proves to have weight."

Some thought, in spite of everything, that what Torricelli had measured was nothing more than the magnitude with which mercury rejects a vacuum. Among these first skeptics was the great French mathematician Blaise Pascal (1623–1662). To put an end to his doubts, Pascal, in 1647, asked his brother-in-law, Florín Périer, to carry Torricelli's barometer to the top of the Puy-de-Dôme (a mountain that sounds all the more familiar to some of us since it is an important summit in the Tour de France). The experiment was carried out the following year. While Pascal's brother-in-law climbed the Puy-de-Dôme with a barometer, another barometer remained in a convent at the foot of the mountain. Since the atmosphere is thinner above the mountains than at sea level, it was reasonable to expect, if Torricelli's mechanical theory was correct, that the mercury would rise less at the summit of the Puy-de-Dôme, as was thus proved. As a good scientist, Pascal then abandoned the animist explanation of the horror of the void and converted to the mechanistic cause.

But the definitive experiment came from the hand of Boyle and the pneumatic machinery that the equally famous English scientist Robert Hooke (1635–1703) had built for him. This machinery operated with a piston and produced the vacuum in a bell jar by extracting the air in successive phases, and the raising and lowering of the piston. Lowering the piston became increasingly difficult; when almost no air remained inside the bell jar, no human force

Figure 2: *Torricelli inventing the barometer. (Illustration from* The Atmosphere: The Great Phenomena of Nature, *by Camille Flammarion. Montaner y Simón, Barcelona, 1902.)*

Figure 3: The real beginning of the barometer (in The Atmosphere: The Great Phenomena of Nature, *by Camille Flammarion. Montaner y Simón, Barcelona, 1902).*

could budge it. This experiment is tantamount to reaching the atmosphere's upper limit, the surface of that ocean of air of which Torricelli spoke.

Boyle placed Torricelli's barometer inside the bell jar and began maneuvering the piston. With each extraction of air, the column of mercury went down, exactly what happened to Pascal's brother-in-law when he climbed to the top of the Puy-de-Dôme with the barometer. Finally, when it was now possible to lower the piston, the column of mercury had almost disappeared. But if any air was allowed into the empty bell jar, the column rose slightly. The ocean of air behaved like that of water, and the air acted as if it were a liquid.

Experiences like these gradually discredited the Aristotelian concept of nature, and little by little scientists stopped thinking in biological terms for interpreting the behavior of inanimate matter and celestial bodies. These behaviors are not comparable to the behavior of living organisms, and they lack any purpose. However, they display distinct irregularities, and the discovery and mathematical description of the laws of physics and chemistry become the objective of these two sciences. But besides that change in thinking, the new natural philosophy introduced precise observation and laboratory experiment as the means for proving a theory or for deciding which, among several possible alternatives, was the true doctrine.

Final Causes

In the cathedral in Burgos, there is a mechanical figure popularly called the *Papamoscas* (the *Flycatcher*), built in the first half of the Middle Ages. It rings a bell every hour and opens its mouth (from which it derives its name). Next to it, a smaller human figure, the *Martinillo* (the *Little Hammer*), sounds on the quarter hour, ringing two bells. If the modern natural scientists wanted to explain nature in mechanical terms, with mechanical figures like the Flycatcher and Little Hammer, they had a perfect example of how to extend that explanation to the living world as well. In Aristotelian science, biology was the model for physics and chemistry. Now the reverse happened: physics and chemistry could be transformed into the model to be applied to biology, an idea some scientists still accept, and are therefore called *reductionists* (since they reduce biology to mere physics and chemistry).

The French mathematician and philosopher René Descartes (1596–1650) thought that animals were the equivalent of mechanical figures, and he tried to explain their operation in purely mechanical terms. The human being was the exception, naturally, because besides having the body of a mechanical figure, it housed a nonmaterial substance, the soul, which felt, thought, and decided what the body was going to do. The problem Descartes faced was figuring out how the body and soul could relate to one another, the one being material and the other not.

As we have seen, what especially encouraged the modern natural philosophers was the exclusion of any notion of finality from their thought; that is to say, the elimination of teleology. The famous astronomer, Johannes Kepler, author of *New Astronomy* in 1609 (a historical book in that it abandoned the model of the planet's circular orbits and substituted in its place that of elliptical orbits), was on the fence between the animist and mechanistic vision of the universe when it came to explaining the cause of the movement of the planets, or the reason for which they revolve around the sun. On one hand, he thought that the cause for the velocity of the planets slowing, to their minimum speed at the point of their most distant orbit from the sun, was owing to the weakness of the power of the planet's *soul,* but on the other hand, he asserted, "When I mentioned that the cause of the movement of the planets diminishes with their distance relative to the sun, I concluded that this cause must be something corporeal"; by *something corporeal,* Kepler was referring to a physical-mathematical type of explanation, like the one Newton later found in the law of gravity.

What kind of explanation was it? Newton limited himself to offering an explanation of the movement of bodies expressed through a law of mathematics: two bodies attract one another with a force that is inversely proportional to the square of the distance that separates them, etc. A constant (G), which is universal, is at work in the formula, meaning that the formula operates independently from whatever the bodies in question may be. The formula could not exist otherwise, but it does. It has to do with, as the later laws of physics and chemistry were developed over time, an explanation of *how* the physical world functions, but no effort is made to know *why* it is one way and not another. Newton

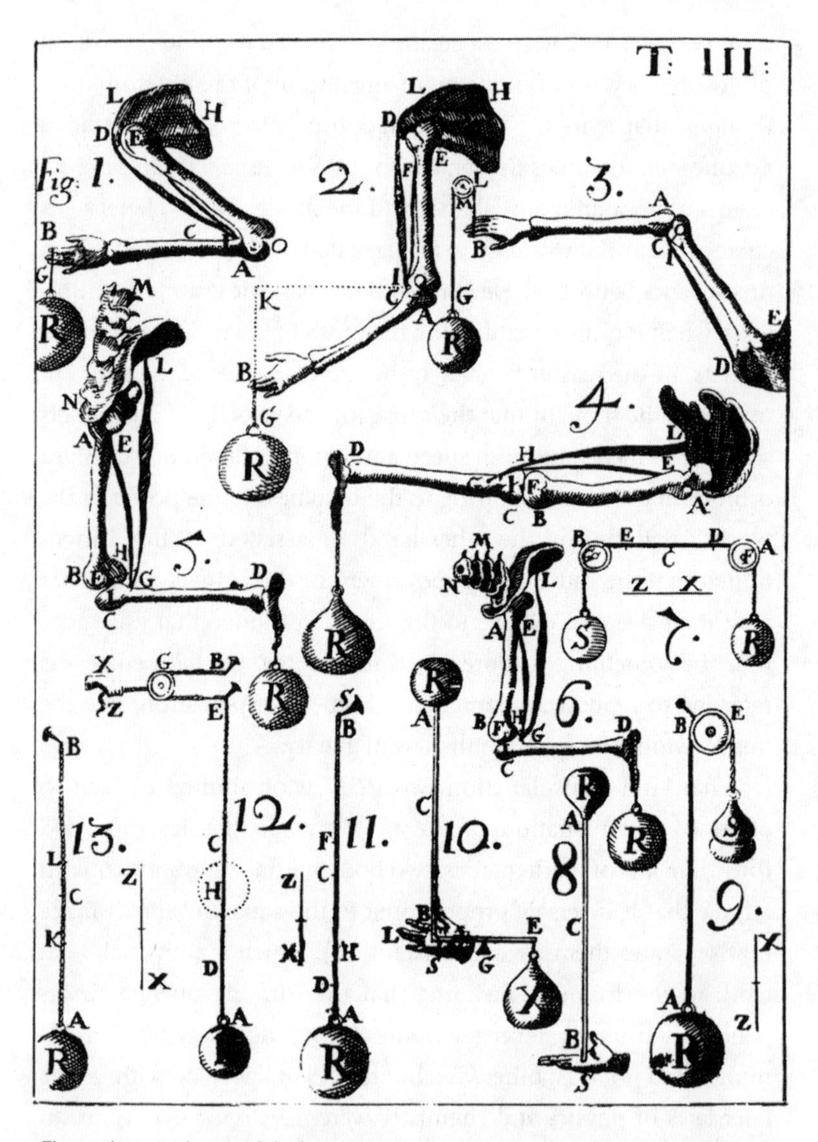

***Figure 4:** Biomechanics of the human arm in an etching by the Italian Giovanni Alfonso Borelli (1608–1679), published in his book* De motu animalium (On the Motion of Animals) *(1680–1681).*

admitted that he had no explanation for the ultimate cause of gravity, and he did not know what its reason was. Quite simply, he was able to formulate it mathematically.

Although, for Descartes, living organisms could be compared to machines, there is a profound gap between their mechanisms, even if it is just a question of a simple bacterium, and those of mechanical figures, like the Flycatcher and Little Hammer in the cathedral at Burgos. In clocks, regardless of how perfect their mechanism, and the same is true for mechanical figures, knowledge resides in the craftsman, whether he be a clockmaker or engineer. In living organisms, knowledge comes from within. All the organisms (and not only animals) that we see altering themselves are *knowledgeable*. That knowledge consists of the capacity to adapt to changes in their surroundings, something which a clock cannot do, and it has a technical name: *homeostasis*. As the environment changes, organisms must continually modify themselves in order to remain unaffected. If, in summer, we expose ourselves to the sun's intense light, *melanisms* (the epithelial cells with the melanistic pigment) make our skin turn dark in order to prevent too much ultraviolet radiation from penetrating the dermis, the layer of tissue located below the epidermis; excessive ultraviolet radiation can produce skin cancer. When winter arrives and we cover up with clothing just as the intensity of the solar light recedes, our darkened skin fades. If we break a bone or suffer a lesion of the flesh, there are natural mechanisms that repair the damage with surprising efficacy (bullfighters, for example, recuperate in a few days from terrible gores that rip open immense gaps in their muscles). If a pathogenic agent or toxin gets into our body, the immune system identifies and neutralizes the agent or toxin. The next time that the same agent or toxin enters the body, the organism *will remember* the previous experience, and the production of antibodies or antitoxins will be that much quicker and efficacious. The success of vaccines remains in the body's immunogenetic *memory*, and in some cases, booster shots are needed from time to time. In animals called *endotherms* (in popular terms, "warmblooded"), the body temperature remains constant within certain limits, despite the changes in atmospheric temperature. The concentration of salt in the blood is also maintained at a constant level with great efficiency by the kidneys. When we live at high elevations, on a big mountain or plateau, where the concentration of oxygen is less than that at sea level, the number of *hematids* (the red blood cells that transport oxygen) increases. The blood thickens because of this, and thus the familiar *hematocrit value* or increase in adrenaline, which has been researched under antipodean controls in competitive sports, since the increase in red blood cells increases the efficiency in transporting oxygen to the muscles (although, of course, with significant risks to one's health).

Like the rest of nature, living organisms can also be studied using physical-chemical methods, and in that way, we will learn *how* they function. The biological discipline that concerns itself with this aspect of living organisms is called physiology. But there is a fundamental difference between living matter studied by biology and inert matter researched by physicists and chemists. The difference is that the notion of finality cannot be excluded from the world of biology because, as we have seen, organisms present us with functions and behaviors clearly oriented toward an objective: they are *intentional*. No one can doubt that living things have a corporeal design that permits them to realize their functions efficiently and remain alive; moreover, in the case of animals, there is a behavior directed toward fulfilling their ends to exist and reproduce. Boyle recognized the necessity of the idea of intentional purpose when it came to living beings: "There is no part of nature, such as we know it, in which the consideration of final causes is as suitable as in the structure of the bodies of animals."

Biology could not, therefore, limit itself to trying to understand *how* the living world functions. It needed to have concrete knowledge of *why* it is one way and not another. In other words, what is it that exists behind the design? Who is its author?

Note: The translation of Lope de Vega's verses is by Naomi Ayala, a published poet and resident of Washington, D.C.

Chapter II

Darwin/Oedipus

The Oracle of Delphi

By the middle of the nineteenth century, the science of biology found itself very much behind the times with respect to physics and chemistry, which had taken a giant leap forward with the seventeenth century's Scientific Revolution. Biology was not advancing conceptually and found itself at an impasse. Rather than a true science, it was a compendium of botanical and zoological curiosities: a natural history that provided information regarding the flora and fauna of different locales; a descriptive intellectual activity, but in no way explanatory. Neither was there a convincing scientific explanation for fossils whose usefulness in ascribing dates and correlating rocks had begun to surface.

The reason for biology's intellectual backwardness was this: no one knew how to explain the *perfection* of living organisms, or in other words, that they are endowed with *intelligent* structures, mechanisms, and behaviors. As we have commented before, organisms present us with *designs* that, aside from amazing us with their complexity, seem to have been created in order to more efficiently perform the activities that allow them to preserve life and propagate it. Those *designs,* therefore, fulfill certain ends; they have objectives. They are like machines we humans build. As Charles Darwin would later say, they are amazingly ideal "regarding the habits of the life" of organisms, which are different in every species. The possible cause of such perfection was God, whose mind, by definition, can neither be known nor be the subject of experimental study. No philosopher had found a better response in all of history. By placing the cause of *perfection* of living things outside of nature, that cause could not then be the object of scientific investigation, since science is concerned only with the material universe. On the other hand, because there are no structures in the world of physics and chemistry that serve specific ends, the problem of *design* did not exist, and so those branches of science were able to progress unabated.

If the *design* in biology obeys a divine *intention,* then it is the responsibility of theologians, and not biologists. But if one excluded God, one was left without an answer to the question of who was the *architect,* what *mind* was it that was able to create so much wonder (and so much beauty)? In fact, as we have already seen, the *perfection* of nature was one of the ways that Saint Thomas had for explaining God's existence.

The situation was similar to that of the city of Thebes in classic Greece. The myth tells of an evil being who arrived at its gates, the Sphinx—a monster with the head of a woman, the chest, legs, and claws of a lion, the tail of a serpent, and the wings of a bird—and who held the Thebans captive in order to ply them with a riddle that it had learned from the Muses, devouring them if they were unable to solve it. An oracle predicted that Thebes would be freed from the monster when someone solved the riddle, which went like this: what creature has no voice, and walks on four legs at dawn, two at midday, and three at sunset? A person named

Oedipus then arrived in Thebes and solved the riddle: it was man, who as a child first crawls, then stands up on his legs, and finally, in old age, leans on a cane. The Sphinx, having gone mad, took its own life, throwing itself off a precipice. Oedipus then received the hand of Jocasta, the widow of the former king of Thebes, Laius, who had been assassinated, under dark circumstances by an unknown traveler, and was made the city's new king.

In the same Greek myth, the city of Thebes lay under siege; biology also lay under siege in the nineteenth century because the enigma of who was the author of the *design* went unsolved. In the same way that Oedipus happened to solve the riddle, Charles Darwin solved the who of the mysterious *architect*. His answer was incredible and astonishing: "No one." That is to say, no individual or being. The author of the *design* is a mechanism of nature, blind and without objectives: natural selection. In other words, the *design* produced itself without the benefit of an *architect* or *intention*. Organisms have structures, mechanisms, and behaviors *with a purpose*, but no one *deliberately* created them.

The similarity between Oedipus and Darwin goes even farther: both suffered terribly because of their resolute search for a truth that, on the other hand, caused them great pain.

The problems in Thebes did not disappear the moment the Sphinx went over the cliff. Once Oedipus became king, a mysterious calamity overtook the city, and Oedipus, who wished to liberate Thebes from the terrible plague, sent his brother-in-law, Creon, to consult with the oracle of Apollo's famous sanctuary in Delphi, as Sophocles (c. 496–406 B.C.) tells us in his famous tragedy *Oedipus The King*:

> Oedipus: Speak it to all; the grief I bear, I bear it more for these than for my own heart.

Creon: I will tell you, then, what I heard from the God. King Phoebus in plain words commanded us to drive out a pollution from our land, pollution grown ingrained within the land; drive it out, said the God, not cherish it, till it's past cure.

Oedipus: What is the rite of purification? How shall it be done?

Creon: By banishing a man, or expiation of blood by blood, since it is murder guilt which holds our city in this destroying storm.

Oedipus: Who is this man whose fate the God pronounces?

Creon: My Lord, before you piloted the state we had a king called Laius.

Oedipus: I know of him by hearsay. I have not seen him.

Creon: The God commanded clearly: let some one punish with force this dead man's murderers.

Oedipus: Where are they in the world? Where would a trace of this old crime be found? It would be hard to guess where.

Creon: The clue is in this land; that which is sought is found; the unheeded thing escapes: so said the God.

Oedipus: Was it at home, or in the country that death came upon him, or in another country traveling?

Creon: He went, he said himself, upon an embassy, but never returned when he set out from home.

Sophocles' work is profoundly tragic because Laius's assassin is Oedipus himself, as his investigations bring him to understand. The more he investigates the matter, the more frightened he becomes, and the closer he comes to the truth, the more afraid he is to face it. He finally realizes that it is he who murdered Laius, who, as if things were not bad enough, was his own father. At the end of the day, Oedipus realizes he has committed patricide, and his anguish knows no limits.

The Question

Charles Robert Darwin was born February 12, 1809, in the English town of Shrewsberry, where his father, Robert Waring Darwin, enjoyed a successful medical practice. Darwin's father was an enormous man, weighing more than 338 pounds and standing six-feet-two inches tall, "the largest man whom I ever saw," according to the son himself. Charles's mother, Susannah, died when he was only eight years old. The naturalist would remember almost nothing about her: "[. . .] it is odd that I can remember hardly anything about her except her death-bed, her black velvet gown, and her curiously constructed work-table" (*Autobiography,* 22). Charles Darwin was never a good student at school, but he always held an interest in nature. In October 1825 (at the age of sixteen), his father sent him to study medicine at Edinburgh University, where his brother Erasmus (five years older) was completing his studies. But neither Charles nor Erasmus felt any inclination to practice medicine, and both abandoned it, Erasmus after one year, Charles the following year. In his *Autobiography*, Charles wrote: "I also attended on two occasions the operating theatre in the hospital at Edinburgh, and saw two very bad operations, one on a child, but I rushed away before they were completed. Nor did I ever attend again, for hardly any inducement would have been strong enough to make me do so; this being long before the blessed days of chloroform. The two cases haunted me for many a long year" (*Autobiography,* 48).

That does not mean Charles wasted his time in the two terms

that he spent at Edinburgh, because, as always, he devoted himself to the natural sciences and made several interesting contacts. One of them was Dr. Grant: "He one day, when we were walking together, burst forth in high admiration of Lamarck and his views on evolution. I listened in silent astonishment, and as far as I can judge, without effect upon my mind. I had previously read the *Zoönomia* of my grandfather, in which similar views are maintained, but without producing any effect on me. Nevertheless it is probable that hearing rather early in life such views maintained and praised may have favoured my upholding them under a different form in my *Origin of the Species*. At this time I admired greatly the *Zoönomia*; but on reading it a second time after an interval of ten or fifteen years, I was much disappointed, the proportion of speculation being so large to the facts given" (*Autobiography*, 49).

Charles's grandfather, also named Erasmus like his older brother, was a physician, naturalist, and poet, and he wrote several works in verse; the most notable being *The Botanical Garden*, *the Temple of Nature* or *The Origin of Society*, and the *Zoönomia* or *The Laws of Organic Life*. *Zoönomia* is where Erasmus Darwin (1731–1802) expounded on his ideas of evolution, which were similar to those of Lamarck in the sense that both thought that organisms were the protagonists in evolution, actively altering themselves in order to adapt more efficiently to their environment.

Erasmus's scientific ideas did not please his son Robert, Charles's father; moreover, Erasmus and Robert did not get along well.

On seeing that Charles would not make a career of medicine at Edinburgh, Robert Darwin sent his son off to Cambridge University to study for the ministry; he spent three academic years there, 1828–1831 (from the age of nineteen to twenty-two). When his father proposed the idea to him, Charles Darwin had his doubts: "I asked for some time to consider, as from what little I had heard and thought on the subject I had scruples about declaring my belief in all the dogmas of the Church of England; though otherwise I liked the thought of being a country clergyman. Accordingly I read with care *Pearson on the Creed* and a few books on divinity; and as I did not then in the least doubt the strict and literal truth of every word in the Bible, I soon persuaded myself that our Creed must be fully accepted. . . .

"Considering how fiercely I have been attacked by the orthodox it seems ludicrous that I once intended to be a clergyman. Nor was this intention and my father's wish ever formally given up, but died a natural death when on leaving Cambridge I joined the *Beagle* as a Naturalist" (*Autobiography*, 57).

As at Edinburgh, the stay at Cambridge proved very beneficial to Darwin's future scientific career, but not to his education as a clergyman. All in all, he passed the compulsory exams: "Again in my last year I worked with some earnestness for my final degree of B.A., and brushed up my Classics together with a little Algebra and Euclid, which later gave me much pleasure, as it did whilst at school. In order to pass the B.A. examination, it was, also, necessary to get up Paley's *Evidences of Christianity*, and his *Moral Philosophy*. This was done in a thorough manner, and I am convinced that I could have written out the whole of the *Evidences* with perfect correctness, but not of course in the clear language of Paley. The logic of this book and as I may add of his *Natural Theology* gave me as much delight as did Euclid. The careful study of these works, without attempting to learn any part of it by rote, was the only part of the Academical Course which, as I then felt and as I still believe, was of the least use to me in the education of my mind. I did not at that time trouble myself about Paley's premises; and taking these on trust I was charmed and convinced by the long line of argumentation" (*Autobiography*, 58–59).

In his *Natural Theology*, William Paley (1743–1805) used the analogy of the clock: the world, as well as the clock, has need of a mechanism. Paley defended the so-called teleological argument, or the argument of a design, to demonstrate the existence of God; thus, he explained the adaptations of living things, which he clearly recognized, as proof of the necessity of an author. Of all the polemics (or ways) used to demonstrate the existence of God, the teleological, or that of a design (or purpose), has always been the most popular, because it was also the most accessible and easiest to understand. It can be summarized with this question: if we cannot conceive of a complex machine, like a clock or bicycle, without an author, and with the idea seeming ridiculous to us that its pieces may have been assembled by pure chance, how could biological mechanisms as perfect as organisms have spontaneously emerged if there is no author? Isn't it obvious that there has to have been an intelligence behind each design at all times? Charles Darwin would dedicate the rest of his life to answering this question, although at the time, in 1831 and at the age of twenty-two, young Darwin still did not know this.

Sir Charles, Naturalist

The experience that marked Darwin's life, in every sense of the word, was his five-year voyage (1831–1836) around the world aboard the HMS *Beagle*. Darwin held a degree in theology and not in science from Cambridge at the time; nevertheless, his interest in nature throughout his life had prepared him for making the most of such an exceptional opportunity. It was on the voyage aboard the *Beagle* that Darwin began pondering the origin of the species, and its adaptations.

Darwin writes the following in his *Autobiography*: "Whilst on board the *Beagle* I was quite orthodox, and I remember being heartily laughed at by several of the officers (though themselves orthodox) for quoting the Bible as an unanswerable authority on some point of morality. I suppose it was the novelty of the argument that amused them" (*Autobiography*, 85).

During his stay in Brazil, Darwin made notes in his diary on the splendor of the Brazilian jungle: "It is impossible to give an adequate idea of the feelings of wonder, admiration, and devotion that fill and elevate my thought." But at the end of his life, in 1876, Darwin would write in his *Autobiography*: "The state of mind which the grand scenes formerly excited in me, and which was intimately connected with a belief in God, did not essentially differ from that which is often called the sense of sublimity; and however difficult it may be to explain the genesis of this sense, it can hardly be advanced as an argument for the existence of God, any more than the powerful though vague and similar feelings stirred by music" (*Autobiography*, 91–92).

On his extended journey through South America, young Darwin discovered other things that made him doubt the immutability of the species.

"During the voyage of the *Beagle* I had been deeply impressed by discovering in the Pampean formation great fossil animals covered with armour like that on existing armadillos; secondly, by the manner in which closely allied animals replace one another in proceeding southwards over the Continent; and thirdly, by the South American character of most of the productions of the Galapagos archipelago, and more especially by the manner in which they differ slightly on each island of the group; none of these islands appearing to be very ancient in a geological sense.

"It was evident that such facts as these, as well as many others, could be explained on the supposition that species gradually become modified; and the subject haunted me. But it was equally evident that neither the action of the surrounding conditions, nor the will of the organisms (especially in the case of plants), could account for the innumerable cases in which organisms of every type are beautifully adapted to their habits of life—for instance, a woodpecker or tree-frog to climb trees, or a seed for dispersal by hooks or plumes. I had always been much struck by such adaptations, and until these could be explained it seemed to me almost useless to endeavour to prove by indirect evidence that species have been modified" (*Autobiography,* 118–119).

With these ideas circling inside his head, Darwin returned home. He married on January 29, 1839 (at the age of thirty) and resided in London until September 1842, when he moved to his permanent residence in Down, in the country, where he would spend the rest of his life. During the three years and eight months of his residence in London, Darwin began suffering the ailments that would accompany him his entire life and caused him to remain a recluse in his house forever: "Few persons can have lived a more retired life than we have done. [. . .] My chief enjoyment and sole employment throughout life has been scientific work; and the excitement from such work makes me for the time forget, or drives quite away, my daily discomfort. I have therefore nothing to record during the rest of my life, except the publication of my several books" (*Autobiography,* 115–116).

The young Darwin, who sailed around the world aboard the *Beagle,* was a quite different person from the mature Darwin, who many years later would be remembered in history for his theory of evolution by natural selection. The image of an old man, with a white beard, of venerable aspect and a reclusive and sedentary life, which one normally associates with him, has nothing to do with that of the active and intrepid naturalist that he was during those five years of his journey. Young Darwin was full of energy, which is evident in the pages of his memorable *Diary of the Voyage of a Naturalist Around the World.* It is an extraordinary account of travels that reads like an adventure novel and is filled with interesting facts about the lands and peoples he came to know. His *Diary* would continue figuring among the classics of travel literature and, it alone, secured Darwin's name in history. That young man of twenty some years of age seemed to let nothing stand in the way of his enthusiasm for learning everything about the countries he visited.

As an example of his temperament, when the *Beagle* docked at Bahía Blanca, Darwin decided not to continue on to Buenos Aires by ship, but to travel by horseback through the four hundred miles of uninhabited territory that separated the two cities. He hired a gaucho to accompany him across the plains. At that time, a relentless war was being waged against the Indians on the Pampas, very similar to what we are accustomed to seeing in westerns, which made the journey very precarious. Darwin set out on September 8. Along the way, he met up with bands of soldiers who were following the route of relay posts. On September 11, they saw a huge cloud of dust being kicked up by a group of riders coming toward them. The horsemen were Indians, recognizable because of their long hair, but fortunately, it was a friendly tribe. On September 15, they passed a bivouac where the Indians had killed five soldiers. The military officer bore eighteen lance wounds on his body. When they reached a post-house on September 19, the owner told them that they could not stay there without a passport because the area was so overrun with thieves that he did not trust anyone. Darwin then showed him his passport, which began with the words "Sir Charles, Naturalist," and he was allowed to stay the night. On September 20, he reached Buenos Aires. A naturalist by nature, he spent the entire trip jotting notes in his diary on every bit of data and a wide variety of topics.

After returning from his trip, Darwin immediately set to work on the transmutations of species. "My first note-book was opened in July 1837. I worked on true Baconian principles [that is to say, following Francis Bacon's inductive method], and without any theory collected facts on a wholesale scale, more especially with respect to domesticated productions [. . .]. I soon perceived that selection was the keystone of man's success in making useful races of animals and plants. But how selection could be applied to organisms living in a state of nature remained for some time a mystery to me.

"In October 1838, that is, fifteen months after I had begun my systematic enquiry, I happened to read for amusement Malthus on *Population,* and being well prepared to appreciate the struggle for existence which everywhere goes on from long-continued observation of the habits of animals and plants, it at once struck me that,

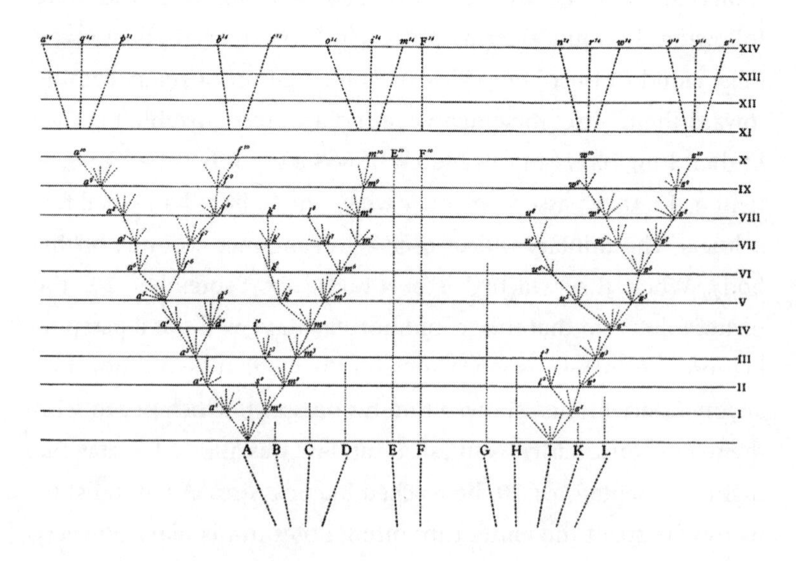

Figure 5: *This is the only diagram contained in Darwin's* The Origin of the Species.

under these circumstances favourable variations would tend to be preserved, and unfavourable ones to be destroyed. The result of this would be the formation of new species" (*Autobiography,* 119–120). The explanation of this mechanism, which he would call *natural selection,* would be the fundamental thread of Darwin's book *The Origin of the Species,* published in 1859.

Darwin would never again believe in the old argumentation of design, which had so profoundly convinced him when, at the age of twenty-two, he read William Paley's *Natural Theology.* "The old argument of design in nature, as given by Paley, which formerly seemed to me so conclusive, fails, now that the law of natural selection has been discovered. We can no longer argue that, for example, the beautiful hinge of a bivalve shell must have been made by an intelligent being, like the hinge of a door by man. There seems to be no more design in the variability of organic beings and in the action of natural selection, than in the course which the wind blows" (*Autobiography,* 87).

Natural selection was responsible for the origin and adaptations of living organisms, including humans, the way Darwin saw it: "As soon as I had become, in the year 1837 or 1838, convinced that species were mutable productions, I could not avoid the belief that man must come under the same law. Accordingly, I collected notes on the subject for my own satisfaction, and not for a long time with any intention of publishing. Although in *The Origin of the Species,* the derivation of any particular species is never discussed, yet I thought it best, in order that no honorable man should accuse me of concealing my views, to add that by the work in question 'light would be thrown on the origin of man and his history.' It would have been useless and injurious to the success of the book to have paraded without giving any evidence my conviction with respect to his origin.

"But when I found that many naturalists fully accepted the doctrine of the evolution of species, it seemed to me advisable to work up such notes as I possessed and to publish a special treatise on the origin of man" (*Autobiography,* 130–131). Here he is making reference to *The Descent of Man,* published in 1871.

Darwin, Parricidal?

What was it that changed an intrepid man, full of vitality, such as the young Darwin aboard the *Beagle,* into a sickly and reclusive person? Did he perhaps contract some grave or incapacitating illness?

It has been postulated that the origin of Darwin's ills was the so-called Chagas' disease (named after the Brazilian physician Carlos Chagas). It has to do with a trepanomiasis, an infection produced by a protozoan, *Trypanosoma cruzi,* transmitted to humans by blood-sucking insects. One of them is the South American vinchuca (*Triatoma infestans*), which Darwin encountered and suffered from on his voyage. In his *Diary of the Voyage of a Naturalist Around the World,* Darwin notes: "I could not rest from having been attacked (I use this word on purpose) by a numerous and bloody group of huge black insects in the Pampas, belonging to the *Benchuca* genus, a species of Reduvius. It is impossible to imagine anything more disagreeable than to feel these soft, wingless insects, about an inch long, crawling over your body."

Chagas' disease is an endemic malady in widespread, rural regions of Central and South America. After a period of acute illness that begins a week after infection, Chagas' disease can turn into a chronic illness that principally causes cardiac arrhythmias. Sometimes death is produced through cardiac failure, when the parasite infests the heart muscle. There is no cure. On one occasion, the famous South African physician and paleontologist

Philip Tobias told us that he thought it probable that Darwin had contracted that disease.

But there is another, even more disturbing, explanation. What Darwin suffered from could also have been a neurosis produced by the shock occasioned by his scientific discoveries, which gradually brought him ever closer to the truth and the paternal figure. Charles Darwin felt a great appreciation toward his enormous father, but also certain resentment. "To my deep mortification my father once said to me, 'You care for nothing but shooting, dogs and, rat-catching, and you will be a disgrace to yourself and all your family.' But my father, who was the kindest man I ever knew, and whose memory I love with all my heart, must have been very angry and somewhat unjust when he used those words" (*Autobiography,* 28).

It is a fact that Charles's father was authoritarian, as his son's ambivalent feelings toward him seem to indicate. The father doubtlessly had disapproved of the theory of evolution upon which Charles was expounding, something that would create conflict in his mind, and moreover, would be aggravated by the very firm and orthodox religious convictions of his dear wife Emma, whom Charles would not want to hurt for anything in the world.

It is curious that Robert Darwin, Charles's father, also had a difficult relationship with his own, no less authoritarian, father, Erasmus, who was an evolutionist like his grandson (and perhaps that is why Robert, in reaction to his father, was not). On the other hand, Charles Darwin had family ancestors worthy of consideration with respect to his possible neurosis. His Uncle Erasmus committed suicide at age forty. His maternal grandfather suffered from bouts of nervous depression, and his mother's brother suffered periods of depression accompanied by strong abdominal problems.

But there are authors who, when using the metaphor of perpetrated parricide (psychologically) for Darwin, are aiming higher.

When Charles Darwin died, April 19, 1882, twenty members of Parliament sent a letter, to the Dean of Westminster Abbey, asking that he be buried next to Newton. The Spanish ambassador attended the ceremony. In their excellent biography of Darwin, Julian Huxley and H. D. B. Kettlewel wrote: "This is how the two major scientists in England's history ended up together: Newton, who had done away with miracles in the physical world and reduced God to the role of a creator of cosmos who, on the day of creation, had set in motion the mechanism of the Universe, subjected to the inevitable laws of nature; and Darwin, who had not only done away with miracles, but also with creation, stripping God of his role as man's creator, and man of his divine origin."

But, in truth, all Darwin did was to carry to the final consequences the program of the baroque scientific revolution, which consisted of explaining natural phenomena as the result of natural causes, without excluding the human being. The fundamental difference between Darwin's biology and Newton's physics is that biology could now answer the questions of *why*? Such as, why do living things exist? Why do humans exist? Why do birds have wings? Or, why do humans possess the faculty of reason? Since Darwin, the answer to these questions is historical, found in evolution, and can be told in the form of a narration that describes how the diverse species including ours, that populate Earth came about, and what shape their adaptations took.

With respect to religion, as his son Francis tells it, Darwin was very reserved and believed this was a personal question that was an individual's private matter. Moreover, Darwin did not wish to wound anyone's sensibility, and on the other hand, he thought that he had nothing to say on a subject on which he had not sufficiently reflected. Consequently, with characteristic humility, Darwin did not consider his opinion on the question of religion to hold any special merit, nor that his scientific authority should extend into that area. In a letter written in 1879, he stated his feelings in very clear terms: "Whatever my personal opinions, the fact is that they are of no consequence to anyone but me. However, because I am asked, I can state that my criteria often varies [. . .]. In my most extreme moments of doubt, I have never been an atheist in the sense of denying the existence of God. I believe, in general terms, (and each time even more so, as I grow older), although not always, that the term agnostic would be the most apt description of my spiritual temperament."

The belief in the existence of God and His role as Creator of the universe is outside the field of science and probably can neither be proved nor disproved based on our knowledge, of the material world no matter how perfect it may become. But nothing that science has discovered is opposed to the believer thinking that Darwin merely completed the task that Galileo had outlined: to read *The Book of Nature*, finding that, in addition to its mathematical terminology, the book was also written in the language of biology.

Billiard Balls

From the time of the publication of *The Origin of the Species* in 1859, the theory of evolution was universally accepted in the scientific world, and biology freed itself from the heavy burden that had impeded its advance. As Theodosius Dobzhansky (1900–1975), a famous evolutionary biologist, has written, nothing makes sense in biology without evolution, to the point that everything discovered before Darwin can be considered irrelevant. But contrary to popular opinion, Darwin's great contribution was not the discovery of evolution, but rather the mechanism of natural selection that explains it—and of which another great scientist was an independent coauthor: Alfred Russel Wallace (1823–1913).

At the end of the nineteenth century, Darwin was praised as one of the great geniuses of humanity, but curiously, the mechanism of natural selection, the real cause of evolution, was barely given any thought. Other mechanisms successfully competed against natural selection.

One of them was *finalism*: evolution guided by God. Although finalism is not *highly regarded* in the academic world of biology, it cannot be said that it has disappeared, in the least, from the minds of many thinkers and popular thought. The fact is that finalism, unlike *Darwinism* (the theory of evolution through natural selection), gives meaning to our presence in this world. One paleontologist and finalist, the French Jesuit Pierre Teilhard de Chardin (1881–1955), was warmly embraced in our country in the 1970s, although today he is barely mentioned within academic circles. But there are scientists who openly express the opinion that the question of whether there is a purpose in evolution, rather than being scientific in nature, is a metaphysical matter, and that deciphering it is outside the purview of science; those who say this, whether they openly state it, are all finalists.

From the materialist camp, there were also writers who maintained that evolution resulted from organisms' internal causes, and not natural selection, which is the action that the environment exercises on organisms, putting them to the test and allowing only the most ideal to survive. In the fossil record, paleontologists saw evolutionary lines that followed straight trajectories, which they explained as the result of evolutionary habits. The environment was too changeable, they said, to produce those lineal evolutions that paleontologists very often detected, and which dominated *evolutionary tendencies*.

Another source of discrepancy with Darwin's *selectionism* was *Lamarckism*. This war an evolutionary theory prior to Darwin's, one that had been proposed by the Frenchman Jean-Baptiste de Monet, Baron of Lamarck (1744–1829). According to Lamarck, adaptation is the consequence of a specific activity developed by an organism: the giraffe, for example, doubtlessly has a very long neck from stretching it so much to eat from the trees' most tender shoots. The descendants, who benefited from them, inherited such changes.

Many learned individuals, who accept evolution as a fact, think in Lamarckian terms without realizing it. It is possible for there to be a planet where modifications produced during its existence are inherited, and thus the most used organs develop and the least used atrophy, but of course, in our world, no such phenomenon occurs. This is owing to the kind of biological inheritance of Earth's inhabitants, an inheritance that resides in the genes, which in turn reside in the chromosomes inside the cells; there is no way for the genes that reside in the sperm and ovum to know which muscles or organs are used most by the individual who produces them.

One day, in the health pages of a Spanish newspaper, we were reading a report on aerospace medicine. A professor of physiology had asked himself, what changes would be produced in man in space in the future, living without gravity? Possibly, in time, he said, after generations, the lower extremities would atrophy until they disappeared, because they would not be needed in an environment where there is no gravity. But what would produce such an evolution? According to the Lamarckian principle, simply the lack of use, not using one's legs. But the explanation, according to the Darwinian mechanism, is quite different; it would claim that very few of the humans born in space would produce progeny; the individuals who did procreate would be precisely those who had shorter legs (although it is difficult to imagine why short-legged astronauts should have more potential for reaching adulthood and producing descendants; if the canons of present-day beauty are preserved, the opposite would happen to them instead, and this consideration is not without merit, because according to Darwin, besides natural selection, by which organisms adapt to their environment, sexual selection is also at work in nature, based on the competition for reproduction). Explained in the language of sports, Lamarckism implies that simply by playing basketball, generation after generation, players would grow increasingly taller because of their efforts reaching for the hoop. Darwinism, on the other hand, defends the idea that the taller an individual is (given his constitution) before beginning to play basketball, the more potential he will have after becoming a star in that sport.

Although *strict* Lamarckism has no defenders, since the changes produced over the life of the phenotype (the physical) are not inherited (are not passed on to the genetic code), its influence persists in the field of evolutionism in a *weakened* form. There are many who think that it is the initiative of organisms, rather than the action of the environment on them, that is the main force behind evolution. While the environment is the main player in Darwinism, it is organisms that play that a role in Lamarckism. This is why Lamarckism is more attractive than Darwinism, because it also gives a certain sense of our origins: we are the result of the efforts of our ancestors; they created us. Important authors like Jacques Monod and Stephen Jay Gould refuse to accept the idea that organisms are mere *billiard balls* in the game of evolution, the clay that natural selection molds. These authors believe that the actions of organisms are what determine the future course of evolution (it is clear that we are talking about behavior that is not genetically programmed, but rather that which spontaneously appears in one or more individuals). In the words of Jacques Monod: "If vertebrate tetrapods appeared and were able to provide the extraordinary growth represented by amphibians, reptiles, birds, and mammals, it is because in the beginning, a primitive fish 'chose' to go exploring the Earth where it was not otherwise possible except for some difficult leaping. It thus created, as a consequence of a modification in its behavior, the drive for selection that must have developed the powerful members of the tetrapods."

Stephen Jay Gould offers a similar version of the same phenomenon: "In a classic and recent case, several species of tits learned to pry open English milk bottles and drink the cream within. One can well imagine a subsequent evolution of bill shape to make the pilferage easier (although it will probably [be] nipped in the bud by paper cartons and a cessation of home delivery). Is this not Lamarckian in the sense that an active, nongenetic, behavioral innovation sets the stage for reinforcing evolution? Doesn't Darwinism think of that environment as a refining fire and organisms as passive entities before it?" (*The Panda's Thumb,* 81).

Kenneth Kardong puts it this way: "The example of the flight of birds also reminds us that a new biological role generally precedes the appearance of a new structure. With a change of roles, the organism experiences new selective pressures in a slightly different niche. [. . .] This initial change in roles exposed the structure to new selective pressures, favoring the mutations that consolidate a structure in its new role. First, the new behavior appears, followed by the new biological role. Finally, a change in structure is established in order to carry out the new activity."

The discussion over what came first, the behavioral change or mutation, is similar to the argument over the chicken and egg, but it is really important when dealing with great evolutionary transitions, the ones that determine the appearance of a type of relatively new organism. Also, in those cases that change the course of evolution, did a behavioral change precede the morphological change? Is there, in the origin of tetrapods, mammals, primates, hominids, or humans, a change in behavior or change in genotype? For George Gaylord Simpson (1902–1984), an important paleontologist, an anatomical change (a *pre-adaptation*) first had to be produced; then individuals equipped with new characteristics could begin to behave differently and occupy a new ecological niche, wherein they live in a different manner.

Another mortal enemy of Darwinism was *saltationism*, the idea that evolution is produced on the basis of great mutations. Natural selection then loses all its importance and ceases to be that patient laborer that continues changing organisms little by little and without pause over a great stretch of time. According to saltationists, the great evolutionary transitions probably took place just once, and natural selection most likely found them already in a finished state. Although saltationism, in its extreme forms, is incompatible with modern genetics, modified versions of saltationism have persisted to this day.

No less favorable to Darwinism were the great discoveries in the new science of genetics in the first quarter of the twentieth century, but in the 1940s, a convergence between geneticists and Darwinists arose, giving birth to a synthesis between the contributions of Gregor Johann Mendel (1822–1884) and Darwin, which was recognized as *neo-Darwinism* or *synthetic theory of evolution*.

Neo-Darwinism is the dominant school in the field of evolutionary thought, but in the last quarter of the twentieth century, there has surfaced among paleontologists (Niles Eldredge and Stephen Jay Gould are its creators) a doctrine called *punctuated equilibrium*, which places in doubt the absolute importance of natural selection in the history of life. In essence, punctuated equilibrium means that while natural selection acts on the members within a species (known as *microevolution*), the history of life (*macroevolution*) is also a consequence of something similar to selection among species. This is because species quickly appear (always in terms of geological time, which involves thousands of years), only to remain relatively unchanged, during which time they behave individually, with their dates of birth and death, and the possibility, which is not always realized, of leaving offspring in the form of a daughter-species. In somewhat schematic terms, species normally do not evolve, and when they do, it is only a small part of the species that changes; that is when a daughter-species appears, as if it were a bud on the branch of a tree. Neo-Darwinism tends to think that species are evolving all the time, because natural selection is a mechanism that never stops. For neo-Darwinism, species are invariably being modified; for punctuated equilibrium, change is reduced to the brief period around the appearance of a species. Neo-Darwinism tends to see the geometry of evolution as essentially linear, while for punctuated equilibrium, ramification is dominant, and out of all the branches, only some of them will produce new shoots, while the majority of species will die out without any descendants.

Years after discovering the mechanism of natural selection, Darwin realized that if a complementary process had not taken place in the past, just one species would exist today on Earth, which would have continued changing throughout time from the moment that life appeared (some 3.8 billion years ago). The history of life is not only about change; it is also about diversification. What does it take for evolutionary lines to branch off, and for species to multiply? Darwin did not have a concrete solution for this problem, and he resolved it by presuming that evolution tended to take on all possible forms of life, all the ecological niches, as one would express it today. In the words of Darwin: "But at that time I overlooked one problem of great importance; and it is astonishing to me, except on the principle of Columbus and his egg, how I could have overlooked it and its solution. This problem is the tendency in organic beings descended from the same stock to diverge in character as they become modified. That they have diverged greatly is obvious from the manner in which species of all kinds can be classed under genera, genera under families, families under suborders, and so forth; and I can remember the very spot in the road, whilst in my carriage, when to my joy the solution occurred to me; and this was long after I had come to Down. The solution, as I believe, is that the modified offspring of all dominant and increasing forms tend to become adapted to many and highly diversified places in the economy of nature" (*Autobiography,* 120–121). Darwin's *principle of divergence,* unlike natural selection, is not a true mechanism, but rather a vague intuition. What does it really do to cause the offspring of species to be successful and multiply, and *tend* to adapt to all possible niches and diversify? The presumed tendencies of nature are never scientific explanations.

It is still not clearly understood how a new species is normally produced from a population of ancestral species. We know that genetic isolation has to appear in some manner; that is to say, it is impossible for members of one population to cross over to those of other populations and produce fertile offspring. There are several possibilities for this, and one wonders which is the most recurrent; this uncertainty can be understood because we, as scientists, have not yet had a period of time in which to directly witness how a species is born (a process that takes thousands of years). At any rate, in most cases, it is believed that the population that was transformed into a new species was geographically isolated beforehand (although, in any case, geographic isolation would not be the cause, but rather a necessary condition).

Since nothing makes sense in biology without the perspective of evolution, everything that has been historically elaborated upon regarding the origin of man, in particular, has been influenced by the different ideas that authors have had on the operation of evolution in general. As an example of this, racist attitudes have been defended on the basis of certain conceptions regarding the evolutionary process.

But the old debates still go on unabated, and the role that natural selection, the cause of evolution for Darwin, has had in our origin and characteristics is still material for controversy. For some scientists, natural selection by *ordinary* means, that is to say, just the way Darwin supposed (and later, the neo-Darwinists) that it functioned, would not be responsible for some of our more important characteristics.

The processes of *heterochronism*, meaning alterations in development, could, according to some, explain our large brain, rounded cranium, and small face, and even, according to others, our biped posture and our inquisitive and investigative nature. This is the theory of *neoteny*, which holds that we are unfinished beings because our development has not been completed; to be somewhat dramatic about it: we are enormous fetuses with a reproductive capacity.

If natural selection is unforgiving and only favors the strongest in the competition among individuals, how is it possible for altruism to have evolved in animal societies, including the human? This is a question that neo-Darwinism must face, and it has done so by bringing to light a specialization called *sociobiology*, which aspires to explain, by means of Darwinian natural selection, social behavior, which includes altruism, cooperative behavior, and limitations on aggression. The creator of sociobiology is the American Edward O. Wilson. The method of this new school consists of adopting into its analysis the *perspective of the gene*, a kind of *gene-centered* view that has been popularized by the Englishman Richard Dawkins as the theory of the *selfish gene*.

Essentially, the explanation of altruism consists of denying it. What seems to us an altruistic act on the part of an individual is really selfish on the genetic level. As an example, because a mother shares half of her genes with each child, there will be two complete copies of her genes in four children, meaning that in the event of an accident, she dies, and her four children survive, her genes become the beneficiaries. From here, the next step is to say, as Dawkins affirms, that genes use the bodies to their own advantage, not "hesitating" to sacrifice them, in that genes are forever and bodies but their mere temporal vehicles. A chicken is merely the instrument that lays an egg, so that another egg might be produced. As will be discussed in more detail later, sociobiology carried into the field of human behavior is today the target of heated criticisms because many see a mistake and danger in any form of determinism being ascribed to human conduct.

We have another important challenge to the importance of natural selection when it comes to the origin of our own mind. Human intelligence probably was not, for an entire group of thinkers (beginning with Alfred Russel Wallace himself), a direct result of natural selection, but rather a collateral or indirect one. Our brain did not evolve in order to read, write, and perform mathematical functions, nor our voice to sing opera, just as our hands did not evolve in order to play the piano; this is obvious, but there are those who say that our brain did not even evolve in order to think (to manipulate symbols), nor our larynx in order to speak (to communicate by means of symbols), but rather natural selection developed these organs for other purposes, even though in our species, they were to abruptly change their function.

We share the idea that the role of natural selection in the history of life deserves to be studied in depth, and that some of the criticisms of Neo-Darwinism, for example, those that come from the field of paleontology, have a sound basis. Natural selection could not constitute the totality of the explanation. But we have no doubt that Darwin, like Oedipus in Greek mythology, liberated biology from the oppression of its particular enigma: how is it possible for there to be a design without a designer?

Notes:

1. The quotation from: *Oedipus The King* in *Greek Tragedies,* vol. 1, trans. David Greene, Chicago: University of Chicago Press. 1960, 114–115.
2. All quotes from: *Charles Darwin, The Autobiography of Charles Darwin 1809–1882,* ed. by Nora Barlow, New York: Norton, 1993.

Chapter III

A Little Bit of Physics and Chemistry

What Moves the World

Perhaps the most prominent characteristic of matter in the universe is *that it moves*. At the very outset of the *big bang*, galaxies were sent scattering at a tremendous velocity. The planets revolve around the stars or other planets. Comets come and go. Earth rotates on its own axis. Oceans endure the ebb and flow of tides, are furrowed by huge currents, and their surface churns with waves. Rivers flow, and the wind blows. Over the course of millions of years, continents shift, are brought together, or separated. Molecules are in constant motion, and the electrons inside them are never at rest.

Living organisms do not remain motionless either. Animals chase one another, search for food, look for a mate, seek shelter . . . Inside their chests, hearts pump rivers of blood, gills move water, lungs move air, and food is pushed along their digestive tracts. None of their organs are at rest. Not even plant life knows repose. Their stems and branches constantly grow, their leaves turn in search for light, flowers open and close, and beneath their surface, their organic fluids continually flow. Microorganisms pulsate, repeating on an infinitesimal scale the activity of macroscopic life. Life also implies movement. Death is but the cessation of movement: eternal rest.

What is it that keeps this universal frenzy active? The answer is simple and, at the same time, very complicated: *energy*. Energy is what moves the world. But what is exactly energy? Physics defines energy as the ability to set a body in motion through space at a fixed speed. Or in simpler terms: energy is the ability to propel or change the movement of things. But this definition does not carry us much beyond what we already know: that energy is what moves the world. This much is clear: although we cannot offer a more satisfactory definition of energy, we are, at least, quite familiar with some of its properties and forms.

Energy presents itself to us in very diverse ways, such as mechanical energy, heat, radiant or luminous energy, electrical energy, nuclear energy, chemical energy, etc. Even though they may appear to be very distinct phenomena, all forms of energy are closely related to one another and can transform themselves into other forms. In order to understand this interrelationship among the different forms of *energies*, it behooves us to take a look at the easiest form of all to understand—mechanical energy which can manifest itself in two ways: as *kinetic* energy or *potential* energy.

Kinetic energy is energy associated with the movement of matter. In other words, it is the ability of a body to dislodge itself in order to move or to reshape another body when the two collide. The concept of kinetic energy is marvelously expressed in an old infantry joke that has a soldier making the comment: "Bullets don't scare me, just their velocity." The apprehensive warrior got it right: kinetic energy, contained in a body that travels through space, is directly proportional to the square of its velocity.

The potential energy of a body is its capacity to acquire kinetic energy with respect to its position in a field of energy. In the area of gravitation, that position is the height at which the body is situated. It is evident that upon being released, objects situated at a greater height attain a greater velocity than those located at a lesser height. The concept of potential energy is also a matter of common sense: "The higher up something is, the harder it falls."

Let's stop and watch the cars that move through our streets. They all possess kinetic energy because they are in motion. This is easy to prove, suffice it to say, if we were to stand in their way. At the moment of impact, they would transfer a part of their kinetic energy to us, knocking us into the air and reshaping (breaking) our bodies. Taking our observations further, we are faced with a mystery: what happens when a car brakes and comes to a stop, at what point has its kinetic energy stopped? Physics assures us, in *the first principle of thermodynamics*, that energy neither creates nor destroys itself. In that case, into what kind of energy has the energy of the car's motion been transformed when the car stopped moving? The answer is *heat*. The brakes and tires of the car, and the pavement, increased the temperature as a result of the friction that has brought the car to a halt.

In fact, heat is also kinetic energy. The molecules that compose bodies are not at rest, but in constant motion. If the body is gaseous, its molecules freely disperse themselves; if it is a liquid, the molecules have a more difficult time; if it is a solid, the molecules can only vibrate, but they are not stagnant. The temperature of a body is merely the sum of the movement of its molecules. If the body (gas, liquid, or solid) is very hot, its molecules move (or vibrate) more slowly. Thus, upon braking, the kinetic energy of the car is converted into the molecular kinetic energy of the brakes, tires, and asphalt.

If we are patient enough and continue observing the car that has come to a stop, we will have the opportunity to confront another enigma: at a precise moment, the car starts off again, once more acquiring kinetic energy. Since we already agreed that energy does not create itself, where did the car get the necessary energy to set itself in motion again? Any child could give us the answer: from *fuel*. The very word sets us on the road to the nature of the energy that moves the car. Fuel means that it can create combustion; that is to say, it can be burned. When combustion is produced, a great quantity of energy is released. This is not difficult to prove either, suffice it to say that when paper burns, its combustion produces light as well as heat.

Energy that is concealed in combustible matter is known as *chemical energy*, and we will return to the subject later.

In any case, a part of the chemical energy of combustible material is transformed into mechanical energy (in other words, energy in motion) in the car's motor, and another part is converted into heat, which makes the motor's temperature rise to the point that it requires continual cooling. In turn, the mechanical energy generated in the motor serves to move the wheels of the car and set it in motion, and also to make the car's generator rotate and produce electrical energy, which is converted into light in the headlamps.

Let us summarize. The primary energy of the car is the fuel's chemical energy. This energy is then transformed into kinetic energy, heat, electrical energy, and luminous energy. But there is still a question that we must answer: what is the origin of the fuel's chemical energy? The answer is simply solar light. How the energy from the sun's light has acted to hold down the carbon deposits of our vehicles is a surprising story, one in which living organisms are the heroes. It is a story that is all the more astonishing if we keep in mind that solar light is also what nourishes almost all living organisms on the planet. But to be able to understand the mystery, and marvel at it, we still need to be better versed in some matters of physics and chemistry.

The Universe Against Us

Although, as we have seen, energy can manifest itself in very diverse forms and can transform some forms of energy into others, we must not think that all types of energy are equal. Using technical jargon, we can make a distinction between *high-grade* and *low-grade* forms of energy.

To help us understand the difference between these two, we will allow ourselves some license with scientific rigidity in order to make an analogy between the concepts of energy and economic value. *The Dictionary of the Royal Academy of the Spanish Language* defines the economic value of an object as "the quality of a given item, for whose worth a certain amount of money, or the equivalent thereof, is given in order to own it." Or, to put it another way: an object's value is its ability to secure for its owner a certain sum of money or the equivalent thereof. If we remember that energy also has the ability to convey movement to matter, we will see that both concepts, economic value and energy, have much in common.

Taking this analogy a step farther, it is clear that economic value can also appear in the form of various objects of merit, such as money, stocks and bonds, consumer goods, real estate, etc. As with forms of energy, these various objects of worth can also be exchanged among themselves. One can have his capital in stocks, cash, gold, diamonds, apartments . . . and exchange one for another. However, we could also make a distinction here between *high-quality* objects of value and *low-quality* objects of value. Those of *high quality* can be exchanged for cash (or their equivalent) in an efficient manner, while those of low quality do not allow for an efficient recovery of the capital (or its equivalent) invested in them. We are not experts in economics, but gold has traditionally been considered a highly valuable commodity, while toothpicks would not seem to be a very good investment. Although the gold and toothpicks have the same original value at the time of investment, the gold maintains its ability to be easily converted into thousands of dollars, or its equivalent, while the same cannot be said of toothpicks. Whereas the value of gold remains strong, the investment of thousands of dollars in toothpicks has been wasted.

This is the key to understanding the concept of forms of high- and low-grade energy. In the former, the capacity to generate movement remains concentrated, while in the latter, the stated ability is dissipated and not advantageous. In the example of the automobile, the fuel is a form of high-grade energy, very useful for producing movement or another form of energy, while the heat generated by the friction from braking is a form of low-grade energy, not very advantageous for producing movement (or another form of energy). On generating heat, the car's kinetic energy dissipated when there was a slight increase in the kinetic energy of the many molecules in the brakes, tires, and asphalt. The gold was turned into a huge supply of toothpicks.

It was the French scientist Nicolas Léonard Sadi Carnot (1796–1832) who discovered that heat is a degraded form of energy, given that it is impossible to convert all the heat generated in any process into movement (technically into *work*). This means that it is not possible to recuperate all the energy used in heating a body in the form of movement, or any other form of energy. From this perspective, heat can be considered a kind of drainpipe for all the other forms of energy.

After this surprising discovery, that heat exists as a degraded form of energy that can neither be used nor transformed, the most discouraging of scientific laws was formulated; it was innocently known as *the second principle of thermodynamics*. This principle, which can be (and has been) stated in very diverse ways, was formulated for the first time by the German physicist and mathematician Rudolf Emanuel Clausius (1822–1888). Clausius' original formulation is really difficult to understand for anyone not versed in thermodynamics, but it can be transcribed into everyday language (and I hope physicists and chemists will excuse us) to something that goes like this: it is impossible to build a refrigerator that, in addition to keeping food cold, generates electrical energy at the same time. Clausius's correct reasoning guarantees us a monthly electric bill.

Joking aside, the most common formulation of the second principle assures us that in a *closed system, entropy* irremissibly tends to increase over time. The concept of entropy signifies the amount of energy that is converted into heat in a closed system. A closed system is that in which neither energy nor matter can enter or exit. Explained in this way, the second principle of thermodynamics does not seem so terrible; but let's put it into terminology that is easier to understand.

Entropy can also be considered a measure of a system's spatial and energetic disorder. The idea of energetic disorder means nothing more than the transfer of high-grade energy to low-grade energy, while spatial disorder, to be precise, is what we normally understand for disorder.

A dramatic example (and the one that interests us the most) of the consequences of the second principle of thermodynamics is the case of living organisms—highly ordered systems of matter. If an organism is transformed into a closed system, depriving it of its source of matter, energy, and food, it dies and decomposes into a heap of highly disordered matter: "Dust thou art . . . "

So in our universe, energy tends to dissipate into a form of heat, and matter struggles not to come apart. What is worse, if the universe, as many scientists assert, is in and of itself a closed system, it then irrevocably tends to fall into disorder. This vision of the universe inevitably falling head first into disorder would justify renaming the universe as the *principle of universal pessimism*: no matter what we do, no matter what we attempt to construct, everything inevitably tends toward chaos; the universe is against us.

However, a ray of light does exist in this quite somber panorama. In 1943, the Nobel Prize winner in physics, the Austrian Erwin Schrödinger (1887–1961), called attention to the fact that the phenomena of life seem to challenge the second principle of thermodynamics, in that, from its origin, living matter gradually acquired at every step an ever greater complexity. This increase of order over time seems to refute the image of a universe condemned to chaos. However, this is merely an apparent contradiction: living organisms are not closed systems. As organisms feed, they are continually acquiring matter and a high caliber of energy. This continual incorporation of energy and matter transforms them into open systems, against which the second principle has no power.

But even though some of us display a tendency to gain weight, living organisms do not seem to accumulate all the high caliber energy and matter they consume. Where then does all that matter and energy end up? What we know is that on a daily basis, we get rid of most of the matter we consume, in the form of carbon dioxide, urine, and sweat. On the other hand, we use a high-caliber energy in generating heat and energy in order to move about and keep our biological machines active; and this energy also ends up getting converted into heat. In other words, we incorporate very ordered matter and a high caliber of energy, and in turn, expel heat and very basic matter. We live in an ordered state in exchange for an increase in the disorder of external matter and energy!

In fact, as living organisms, we maintain our extraordinary level of order thanks to the constant flow of matter and energy that passes through us, adding to their disorder in the process. Technically speaking, we hold our entropy in check at the expense of increasing that of our environment. The amount of disorder we generate is greater than that of the order we acquire (consider, for example, the amount of heat and waste our cities produce).

In fact, as living organisms, we are not in violation of the second principle of thermodynamics, since on generating a greater external disorder than our internal order, we increase the total disorder of the universe (its entropy). We are merely accelerators of chaos. This is the price of existence.

But how does this process, so easily described in general terms, take place? What kind of high-caliber energy does living matter use? How does it transform that energy into heat and life?

Us, the Sheep, and the Pasture

Shepherds who lead their flocks along the skirts of the Sierra de Atapuerca range are observant individuals, and their conversation is often enriching. One summer afternoon as we were returning from the digs, we decided to stop next to a spring where water flows from the inner recesses of the Sierra. There we met up with one of the shepherds, an acquaintance of ours. We greeted one another and began a casual conversation while absent-mindedly observing the flock drinking. "It's strange," our friend suddenly said, "that sheep and humans drink from the same water." Before such banality, we looked sideways at each other and started to say our good-byes. But shepherds are not used to having many occasions to chat with someone about questions they have, and so, ignoring our apparent rush to be off, our friend continued: "Humans can't eat sheep because we're all animals and made of flesh and blood, but how can sheep just eat grass? Grass isn't meat . . . " He stopped for a moment, then immediately resumed his reflections: "It seems to me that even though grass and meat look different, they must be made of the same thing." We looked at each other again, but this time disconcerted by the shepherd's acumen. "But the strangest thing of all is that the grass grows out of the earth," he continued, "as if both of them were made from the same thing." He stopped again, whistled for his dog to round up some sheep that were straying, and then fixedly looked at us and told us what had dawned on him: "For me, the earth, grass, sheep, and all of us are made from the same thing, don't you think so?" Astonished, we were at a loss for words.

Without knowing it, our friend had followed a path similar to the one taken by the Roman philosopher and poet Titus Lucretius Carus (c. 99–55 B.C.) in his book *On the Nature of Things* in arguing that the universe's physical matter was made up of very small and indivisible particles (*atoms*, in Greek) all made from the same type of substance, differing from one another only in shape and size:

> [. . .] And the smiling meadows are become
> Herds of animals, and into our very bodies
> Are animals of the field transformed, and often
> With our bodies does the mettle
> Of predators and carnivorous birds increase.

For his part, Lucretius merely elaborated on the ideas of the Greek philosopher, mathematician, and physicist Democritus (c. 460–380 B.C.) and those of his disciple, the philosopher Epicurius (c. 341–270 B.C.). According to Democritus, all the bodies of the universe were formed by different aggregations of these particles. What made one object different from another was simply the number, shape, and weight of its atoms. Apparently, the idea did not originate with Democritus, although it was he, rather than his teacher, the philosopher Leucippus (c. 460–370 B.C.) who developed it. Democritus's theory included other aspects, such as the one that says that the spirit is composed of atoms of a special nature, which are shared by the whole body.

The central idea of Democritus's theory (and that of Lucretius), that all the matter in the universe is composed of aggregations of infinitely small particles, composed of the same substance, is basically born in light of our current knowledge of the precise structure of matter. The fact that our shepherd friend should reach, of his own accord, conclusions so perceptible and similar to those of the two philosophers, who lived more than two thousand years ago, justifies our astonishment.

The modern version of the atomic theory was begun by the British scientist John Dalton (1766–1844), who, in 1807, suggested that the universe's matter was comprised of a group of atoms, as well as chemical elements, which differed among themselves in mass, but not in the substance of their composition.

Today we know precisely the exact size of these atoms, which is really small. Hydrogen, the smallest atom, has a diameter of 0.37×10^{-7} millimeters. In other words, nearly 4 million hydrogen atoms lined up together fit inside a millimeter. Atoms are not very heavy either; it takes less than a trillion atoms (in fact, only 602,214 billion atoms are needed) of hydrogen to obtain a weight of one gram.

Until the end of the nineteenth century, we did not really begin to understand the precise structure of atoms. But the first thing we discovered is that the choice of a name was not quite right. As we have already said, atom means indivisible in Greek, but today we know that what we call an atom is, in turn, composed of even smaller particles of matter. These particles are generically called *elemental particles*, and at the present time, more than two hundred different kinds are known.

However, and fortunately for the purposes of this book, only two kinds of particles are responsible for the chemical properties of atoms, and they are the ones that determine the way in which some atoms are bonded together to form larger aggregates—the molecules with which we will later concern ourselves.

The particles to which we are referring are called *electrons* and *protons*. Electrons were discovered by the Englishman Joseph John Thompson (1856–1940), Nobel Prize in physics 1897, and they are so small that it takes about 1,838 electrons to equal the weight of one atom of hydrogen. But the most important property of electrons is not their size, but the fact that they produce a negative electrical charge. Electrons are very important in our daily life; their transferal through a conductor constitutes the electrical current that feeds electricity to our homes, and the images on our television screens originate from a beam of electrons.

Protons were discovered twenty-two years later following an experiment headed by New Zealand–born British physicist and winner of the Nobel Prize for physics, Ernest Rutherford, of Nelson (1871–1937). In his experiment, nitrogen atoms were exposed to radioactivity (to be exact, to a bombardment of *alpha* particles). As a result, he obtained the following: oxygen atoms and some particles—almost 2 million times heavier than electrons—with a positive electrical charge: protons.

Besides electrons and protons, which are responsible for the chemical properties of atoms, there is a third particle of great interest if you are to understand atomic structure: these are *neutrons*, discovered in 1932 by another British physicist, also a recipient of the Nobel Prize for physics, James Chadwick (1891–1974). These particles produce a mass only slightly greater than that of protons, but just as its name indicates, they have no electrical charge. A neutron is formed by joining an electron with a proton[1] (and a third particle, with no electrical charge, and which probably has no mass, called a *neutrino*). Given that at the heart of the neutron, which is an electrically neutral particle, one finds an electron (which has a negative electrical charge) and proton (which has a positive electrical charge), it is easy to deduce that the electron and proton charges are of the same magnitude, even though of contrary symbols. This fact is fundamental to understanding the chemical properties of atoms.

The first step toward an understanding of precise atomic structure was taken by Rutherford, who established two fundamental aspects. In the first place, most of an atom's mass is concentrated in a region of space with a positive electrical charge situated in its center: the *nucleus*. He also suggested that electrons freely collide as they traced orbits around this nucleus. Because electrons have a negative charge, Rutherford added another idea to his model: the nucleus must produce a positive charge of the same magnitude as the sum of the negative charges of its electrons.

However, the way in which electrons are distributed around the nucleus was not correctly explained by Rutherford's model, but by a development, between 1913 and 1915, by the Danish physicist Niels Henrik David Bohr (1885–1962), for which he received the 1922 Nobel Prize in physics.

Bohr's achievement was to integrate into his model the *quantum* theory, proposed by the German physicist and Nobel Prize winner

1. In fact, the neutron is not formed by the joining of an electron and proton. Nevertheless, when a neutron is split, a proton and electron are released, so that it could be said, more fittingly, that it is *as if it were formed by a proton and electron.*

for physics, Max Karl Ernst Ludwig Planck (1858–1947). According to Planck, matter does not emit or absorb energy in a continuous stream, but rather in the form of small, discreet units of energy. Planck had a name for these minimal *packages* of energy: *quanta*. Depending on the type of energy emitted or absorbed, the amount of energy of the quanta is either greater or lesser. Blue light, for example, is composed of quanta of energy greater than those of red light, and the quanta of ultraviolet light are of an even greater energy than those of blue light.

But let's return to the positioning of electrons. According to Bohr's model, electrons cannot occupy just any spatial position in their movement around the nucleus, but instead are tightly arranged in a series of fixed orbits around the nucleus. Each one of these orbits or shells has a distinct level of energy, and the difference in energy between two contiguous orbits is that of a quantum. In effect, this difference continues growing ever larger as we move away from the nucleus. In other words, between shells 1 and 2 there is a leap of energy that is smaller than the existing energy between shells 2 and 3, which in turn, is smaller than the existing energy between shells 3 and 4, and so on, in succession.

For an electron to be able to pass from one layer (shell) to the next layer of greater energy, it must absorb a small quantity of energy (Planck's quantum), which is released if the electron makes the inverse movement and returns to the layer (orbit) of lesser energy. This release of energy takes place in the form of *electromagnetic radiation* (light is electromagnetic radiation). A quantum of energy in the form of electromagnetic radiation is known as a *photon*, so it can also be said that the electron emits a photon when it moves toward a layer of lower energy.

Obviously, the orbits among which the electron moves will release a greater amount of energy the farther removed they are from the nucleus. So that, if an electron from the shells that are most distant from the nucleus moves to an orbit of lower energy, it will release a high-energy photon, of the type known as ultraviolet light, while an electron positioned in an orbit of low energy (that is, closer to the nucleus), which moves toward another orbit even closer to the nucleus, will emit a low-energy photon, of the type known as infrared light. Electrons of intermediate orbits emit photons in the spectrum of visible light. In a similar manner, electrons absorb photons in order to move from low-energy orbits toward high-energy orbits. Low-energy photons can only produce movement in electrons in orbits closest to the nucleus, while high-energy protons facilitate the movement of electrons positioned in orbits most distant from the nucleus.

This concept is so important to understanding how living organisms function that it deserves the time to attempt to explain it by analogy. Imagine a stairway with progressively higher steps (which would be the equivalent to the levels of energy, or shells, in which the electron can move around the nucleus). Now imagine that we wish to place ourselves (playing the role of an electron) on one of those steps. Obviously, we will use a specific amount of energy, which we could call a *unit*, each time we climb one step, so that we can say that climbing the first one costs us one *unit* of energy, then expending another *unit* in going from the first to the second step, and so on, in succession step after step. These *units* in our example are the equivalent of Planck's quanta. Just as each step rises higher than the previous one, each *unit* we use (on going up higher) is lengthier than the one that precedes it. That is to say, there are *units* lengthier than others (in the same way that there are quanta with greater energy than others). If we now choose to descend the stairs, we will release energy corresponding to one *unit* with each step we take, and this energy is converted into sound with every footstep that lands on the stairway (in fact, the energy we release with each footstep in our descent is converted into sound when our feet strike the stairs and into heat when our bodies create friction against the air, but let's admit that all the energy created by our footsteps is also converted into noise each time we strike the step below). The sound that releases the energy of one *footstep* between the steps on the stairway would be the equivalent, in our example, of the photon emitted by an electron in its movement to a layer of lower energy. Because the steps of the stairway are positioned at different levels, the noise produced in our descent would be greatest on the highest steps.

Similarly, an atom can emit photons of greater or lesser energy depending on the difference in energy that exists between the layers through which the electron moves. Just as we cannot remain in a state of suspension between two steps on the stairs, nor take half *steps*, neither can electrons take up positions in between the different rings, nor absorb or admit a quantity of energy lesser than a quantum (or photon, if it is in the form of electromagnetic radiation).

Although it is difficult to believe, this type of energy, with electrons enclosed around the atomic nucleus and released when they move to lower levels of energy, is what keeps us alive, preventing the second principle of thermodynamics from doing us in.

I'm Okay, You're Okay

May 8, 1794, was a fateful date for France and science. That day, the guillotine cut short the life of one of the greatest scientists in history, the father of modern chemistry, Antoine-Laurent de Lavoisier (born in 1743). Among his many achievements was the discovery that the quantity of matter remains constant in chemical reactions. He decisively contributed to the discovery of *oxygen* (which he named and means *generator of acids*) and confirmed the findings, realized at the same time (although independently) by British scientists Henry Cavendish (1731–1810) and James Watt (1736–1819), that water is formed by the combination of two different substances: oxygen and another gas, which Lavoisier named *hydrogen* (*water generator*). We also owe him recognition for demonstrating that the process known as combustion is simply the combination of oxygen and combustible substances (or *oxidation*), and for identifying animal respiration as a form of oxidation.

Lavoisier formalized the concept of a *chemical element* as that of a simple substance, which cannot decompose into other simpler substances. He also proposed, very skillfully, that the combination of these elements in multiple compounds was the origin of all the other substances in the universe.

As we have already seen in the previous section, Dalton departed from this concept of *element* in order to propose his atomic theory of matter. Dalton believed that each element was composed of a single class of atoms, which were equal among themselves, but distinct from the atoms of another element. What accounted for the difference among the atoms of the different elements was not that they were composed of different substances, but rather they had unequal masses. That is to say, a lead atom was not composed of a substance different from that of a hydrogen atom; simply it was denser. Dalton was half right: all atoms are composed of the same type of matter (here he was right), electrons, protons, and neutrons, but they are not distinguished by their distinct quantity of mass (here he was wrong), but rather by the number of protons, neutrons, and electrons that make up their composition.

Let's remember that the protons and neutrons in an atom are found in the nucleus, and the electrons around the periphery.

However, the real difference among the different chemical elements lies *in the number of protons in their nuclei*, a discovery attributable to the work of British researcher Henry Gwyn Jeffreys Moseley (1887–1915). Thus, the simplest element, hydrogen, only has a single proton in its nucleus, helium two, lithium three, and so on, in succession, up to the most complex element known, meitnerium, with its 109 protons.

We also had established that atoms contain, under normal conditions, an equal number of electrons and protons. Thus, hydrogen has one electron, helium two, and lithium three, all the way up to meitnerium with its 109 electrons. However, while protons are firmly set in the nucleus and cannot be extracted from it unless the atom is split (technically, through fission), electrons, because they are on the periphery, can either be added or lost. When an atom gains an electron (with its negative electrical charge), the electrical equilibrium between protons and electrons is broken, and the atom remains negatively charged, while inversely, if it loses an electron, it will acquire a positive charge. In both cases, the atom is ionized; that is, it is converted into either a negative *ion* (anion) or positive *ion* (cation). The ions of an atom retain their composition as the same type of chemical element as the original atom. Soon we will see how to make an atom either lose or gain electrons. But first, let's turn our attention to neutrons.

The quantity of neutrons that can exist in the nucleus of an atom is variable, unless it is transformed into another element because of it; in other words, unless they change their chemical properties. Hydrogen's most common atom has no neutrons in its nucleus, but there are hydrogen atoms that have one neutron, and it is possible to acquire, by artificial means, a variety of hydrogen with two neutrons. Depending on their number of neutrons, the different variants of the atoms of an element are called *isotopes*. The hydrogen isotope with a single neutron is given the name *deuterium*, and the artificial isotope of two neutrons is called *tritium*. But let's not forget that *normal* hydrogen (without any neutrons), as well as deuterium and tritium, are still hydrogen atoms and have the same chemical properties.

Let's return now to the case of electrons. Why should an atom either gain or lose an electron, and thus acquire either a positive or negative electrical charge? The answer is found in the particular manner in which electrons are situated around the nucleus of the atom.

As we have already seen, Bohr's atomic model assures us that electrons are distributed in orbits around the nucleus.[2] But not all layers accommodate the same number of electrons. The number of electrons in successive layers continues increasing according to rule $2n^2$, where n is the number ascribed to the layer. For example, only two electrons *fit* in the first orbit, a maximum of eight electrons is possible in the second orbit, eighteen electrons in the third, thirty-two in the fourth, and fifty, seventy-two, and ninety-eight electrons in orbits five, six, and seven respectively.

Electrons follow a very complex order at the time they arrange themselves in the different shells. Even though you may not remember it, you probably studied this in college and surely passed the required exam. But do not fear; it is not necessary to know the precise order to understand why some atoms tend to lose electrons, while others *prefer* acquiring them. It suffices to remember one rule: atoms reach their maximum stability when they have *eight* outer electrons. This rule has one exception in the

2. In fact, electrons are not distributed around the nucleus in concentric layers, as Bohr thought, but instead have a much more complex spatial distribution, which was later described by Schrödinger. Based on this model, electrons are not arranged in stable orbits or shells around the nucleus, but are found with greater probability in fixed regions of the space, in a very complex and variable form, around the nucleus. Nevertheless, in order to understand the problems dealt with in this book, we can give ourselves the license to imagine layers concentrically arranged around the nucleus.

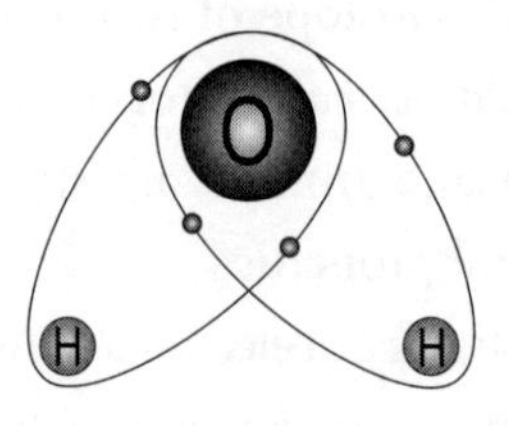

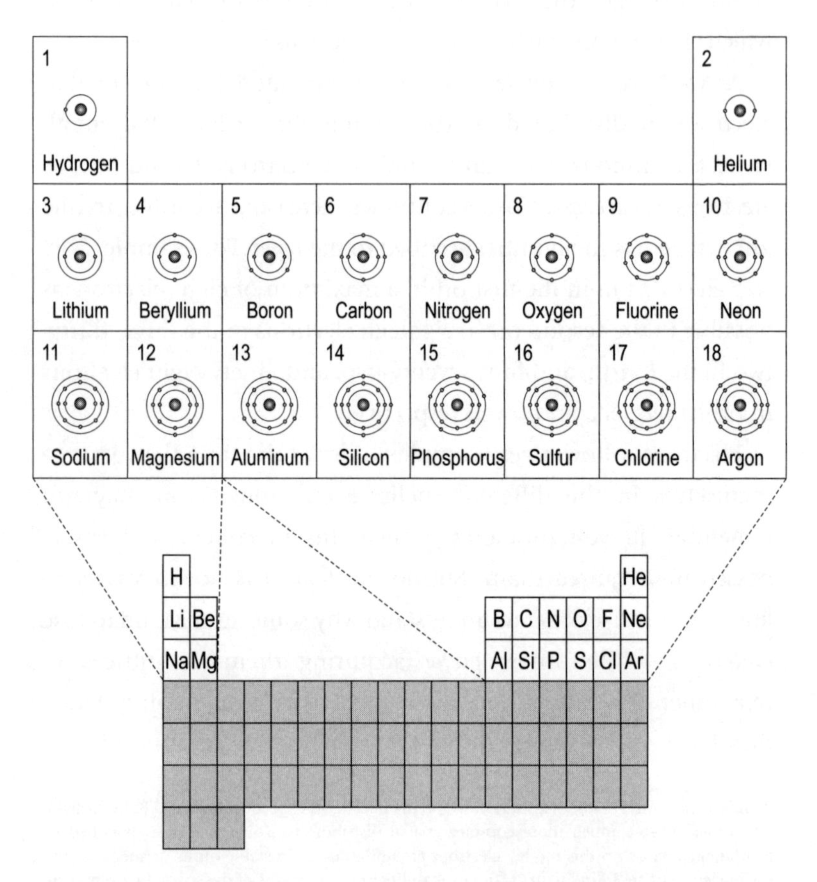

Figure 6: *Above: A molecule of water with its two molecular orbitals. Each of them contains one oxygen electron and one hydrogen electron. Below: The eighteen primary elements of the periodic table showing the placement of their electrons in the different shells.*

case of helium, which only has two electrons in its one and only layer. But because only two electrons fit there, the helium atom's outer layer is full and thus quite stable.

The elements whose atoms have eight electrons in their outer layer are: *neon, argon, krypton, xenon,* and *radon*. Their names are not completely unknown to us; neon, argon, and krypton, for example, are used in the manufacture of fluorescent tubes and light bulbs . . . and yes, Superman's native planet is also called Krypton.

In order to continue moving ahead, you should keep a *periodic table* at hand; this is a systematic list of the different elements, whose original formulation we owe to the Russian chemist Dmitry Ivanovich Mendeleyev (1834–1907). On the periodic table, the elements are arranged, in increasing order, according to the number of protons in their nuclei, so that hydrogen is the first and *meitnerium* the last. But it is not a question of a simple alignment of elements; the 109 elements are arranged in eighteen columns and nine rows of varying lengths. This apparently capricious arrangement of the elements has those with the same number of outermost electrons occupying the same column.

Before proceeding, it behooves us to stop a moment and simplify the explanation. In order to be able to understand the fundamental chemical questions on which life is based, we do not need to understand the properties of all the elements of the periodic table; those of the first eighteen elements, found in the first three rows, do quite nicely.

The elements in the helium column are characterized by having eight electrons in their outermost shell, plus helium with two electrons in its one layer. We have already seen that the atoms of these elements have great stability; which is to say, they are not *disposed* to gaining or giving up any electrons.

The column headed by *fluorine* and *chlorine* contains elements with seven outer electrons. This situation is not *acceptable* to the atoms, because they need eight outer electrons in order to acquire stability. The most direct way to achieve this is to capture an extra electron and thus join together eight outer electrons. Unfortunately for them, when accepting an additional electron, the number of negative charges (electrons) surpasses the number of positive charges (protons), and the atom is negatively charged (let's remember, is converted into an ion). In effect, achieving the stability that eight outer electrons confer on the external ring implies the loss of the atom's electric stability.

Oxygen and *sulfur* are the elements that interest us in the column displaying atoms with six outer electrons. In order to acquire eight electrons in their outer sphere, they must be able to capture two additional electrons, but this implies a great electrical instability, since the atoms are charged with a double negative charge. This problem becomes more acute in the column with *nitrogen* and *phosphorous*, because by having five outer electrons, they would need three more to achieve eight, and gaining those three electrons would produce a net sum of three negative charges for the atoms.

The column with *carbon* and *silicon* represents a dilemma for those elements that possess four outer electrons. A double possibility presents itself here: either to gain four electrons, which would give them a total of eight outer electrons or to lose the four electrons of their external shell, thus transforming their penultimate shell into their outer most layer. Carbon only has two rings of electrons, so if it were to lose the four from its external ring, it would achieve a situation identical to that of helium: two electrons in a single orbit. Silicon has three shells (one more than carbon), so the loss of the four electrons would transform its second layer, where it has eight electrons, into the outermost layer. The first alternative (gaining four electrons) would cause a surplus, with the corresponding negative charge in the atoms of these elements, while the second possibility (relinquishing four electrons) would effect a net sum of four positive charges. Both situations are impossible under normal conditions.

The column that includes *boron* and *aluminum* is characterized by elements having three external electrons. If they were to lose these three electrons, boron (which has only two shells) would attain the configuration of helium (two electrons in a single orbit), and aluminum would be left with eight electrons in its external orbit. In both cases, they would be charged with three positive charges.

The *beryllium* and *magnesium* atoms have two electrons in their external orbits . . . and *are wishing* to lose them. In achieving this, beryllium *would succeed* in finding itself with a single shell with two electrons (like helium), and in the case of magnesium, the penultimate shell, which has eight electrons, would become the outermost shell. In this situation, the respective atoms would be left with two positive charges.

Finally, we encounter *hydrogen, lithium,* and *sodium,* all with a single electron in their external orbits. The loss of this electron would leave sodium with eight outer electrons, lithium with the same configuration as helium, and hydrogen without any electron. It is relatively easy for these elements to be able to lose the *surplus* electron, even though they would all be left with a positive charge.

As for hydrogen, it deserves an additional comment. In theory, it too could gain an electron and complete its single layer in order to acquire the same status as helium, but the sole proton in its nucleus is not strong enough to realize such a feat.

Let's see what we have culled from all this confusion about electrons and orbits. In the first place, there are elements (fluorine, chlorine, oxygen, sulfur, nitrogen, and phosphorous) whose atoms have a tendency to add electrons in order to reach a more stable configuration. Such a tendency is called *electronegativity.* On the other hand, another group of atoms (hydrogen, lithium, sodium, beryllium, magnesium, boron, and aluminum) show a propensity for relinquishing electrons (*electropositivity*). Finally, there is a third group, that of carbon and silicon, both of them *indecisive* between the two tendencies.

It is important to emphasize that some of these atoms cannot satisfy their *wishes* to add or lose electrons. It is relatively easy to add or lose one electron; accomplishing that with two is quite difficult, and it is nearly impossible to remove or capture two or more electrons.

When an atom that is prepared to add an electron is found with another one that *wishes* to lose an electron, an electrical transference from the second to the first is realized. But on carrying it out, the donor of the electron acquires a positive charge, while the recipient is left with a negative charge. Since opposite charges attract, both atoms are brought together by the force of the electrical attraction. This type of union is called an *ionic bond,* and the compounds formed in this fashion are *salts.*

Here is how our table salt is made: one chlorine atom acquires one electron of the sodium atom. The chlorine remains negatively charged and the sodium positively. The electrical attraction forces them to unite in a compound called *sodium chloride.*

The ionic bond helps some atoms acquire the stability they so greatly *crave,* because it allows the electrical charges, which appear when the atoms relinquish or add the electrons they need for stability, to be neutralized. But we have already seen that this situation is not feasible for all atoms, because not all of them can surrender or gain as many electrons as they need. On the other hand, the ionic bond can be broken by water (salt easily dissolves in water), and so it does not seem to be a very long-lasting solution.

Fortunately for atoms, there is another possibility: the *covalent bond*. One atom of oxygen and two atoms of hydrogen strongly bound together form a molecule of water. What is it that holds these atoms together? Remember, perhaps, that oxygen has six outer electrons, so it needs two more for a total of eight to attain stability. On the other hand, hydrogen only has one outer electron. The easiest solution for the problem of oxygen and hydrogen could be for two hydrogen atoms to relinquish their electrons to one atom of oxygen. This situation would have each atom of hydrogen acquiring a positive charge and oxygen two negative charges. The electrical attraction would link the two hydrogen atoms with the one atom of oxygen, with the charges compensating each other. The idea is a good one, but impossible, simply because the oxygen atom cannot be charged by two additional electrons.

In that case, how do the two hydrogen atoms and one oxygen atom remain bonded? Since the transference of electrons between them is not possible, there is another solution: *sharing electrons*. Between each hydrogen and oxygen atom, an integrated molecular orbital is formed, each of them by the electron from one hydrogen atom and another from the oxygen atom. These electrons, shared and located in each molecular orbital, no longer move around either the hydrogen atom or oxygen atom, but around both. The oxygen atom now has four outer electrons plus four others (two of its own and another two contributed by the two hydrogen atoms) in the two molecular orbitals shared with each hydrogen atom. Four plus four gives us eight. At the same time, each hydrogen atom has two electrons in its molecular orbit that are shared with the oxygen atom. Two electrons in a single orbit is the configuration of helium.

Through the process of sharing electrons, atoms are able to attain stability in pairs. But as happens with other pairs, the situation is not always one of equality. In the case of water, the oxygen atom attracts electrons with greater force than the hydrogen atom, the result being that the electrons spend more time near the oxygen atom than its molecular *mate*. As a consequence, the molecule of water shows a little bit of a negative charge next to the oxygen atom and somewhat of a positive charge alongside each hydrogen atom. All in all, they are better off together than apart.

The covalent bond is very strong (the diamond, the hardest known substance, is formed by carbon atoms tightly bonded together by covalent bonds), and moreover, it does not dissolve in water like the ionic bond. This last property is very useful in forming bodies that, like those of living organisms, are comprised for the most part of the liquid element.

For each electron that an element needs to gain or lose, it can form a molecular orbit (and therefore, a covalent bond) with another member disposed to sharing electrons. Thus, the hydrogen atom can form a covalent bond, and the oxygen atom and its column mate in the periodic table, sulfur, can form two covalent bonds. The nitrogen and sulfur atoms can combine for three, while the carbon and silicon atoms hold the record for covalent bonds, being able to form up to four. These different capabilities for establishing covalent bonds make it possible for some atoms to form larger molecules than others, and it explains their distinct importance in the molecular structure of living things.

Life's Building Blocks

As we have seen thus far, our shepherd friend was correct when he asserted that the land, grass, sheep, and all of us are made *of the same thing*. Not knowing chemistry, our friend does not know that that *thing*, of which we are all made, consists of the atoms of the different chemical elements. However, on the one hand, the proportions in which the different elements are found in the earth are not equal, and on the other hand, neither are they of equal proportions in grass, sheep, and humans.

Different combinations of barely eight elements form 98 percent of Earth's crust. Among them, the most abundant by far are oxygen (47 percent) and silicon (28 percent). The next element in importance is aluminum (8 percent), which is followed by: iron (4.5 percent), calcium (3.5 percent), sodium (2.5 percent), potassium (2.5 percent), and magnesium (2 percent). As for hydrogen (0.2 percent) and carbon (0.2 percent), just minimal traces are found in Earth's crust.

On the other hand, living organisms are almost exclusively formed by three elements. If we take ourselves as an example, our bodies are composed of 63 percent hydrogen, 25.5 percent oxygen, and 9.5 percent carbon. All together, these three elements make up 98 percent of the total number of atoms of our organism. If we add nitrogen to them (1.4 percent), we now know 99.4 percent of our chemical composition with respect to the elements. Surprisingly, more than 70 percent of the body's atoms consists of hydrogen and carbon, two elements that, as we have already seen, are scarce in Earth's crust. In short, hydrogen and carbon are more abundant in us than in rocks, 286 and 50 times more, respectively. It is also noteworthy that we have half the amount of oxygen and a great deal less silicon than the soil in which the grass grows. So that we, as living organisms, are not mere fragments of Earth's crust.

However, you could be thinking that we have erred in our comparison. When all is said and done, who said we have to have the same composition as Earth's crust? Since life originated in the sea, the logical thing is to think that our chemical composition is similar to that of oceans, not of rocks. In effect, our bodily composition is very similar to that of the ocean in one aspect: oceans, as well as living organisms, are principally constituted of water (between 70 percent and 90 percent of our matter is water). But the comparisons end here. In one liter of ocean water, there are, on average, 35 grams of other dissolved chemical elements, and around 90 percent of this quantity consists of chlorine (55 percent), sodium (30.5 percent), and magnesium (3.7 percent); there is barely any carbon in seawater (less than 0.01 percent). Therefore, even when we compare it to the medium in which life originated, the chemical composition of living organisms is still quite different.

There is one curiosity regarding carbon. We, as living organisms, act as veritable intake valves for this element on our planet. The amount of carbon contained in living beings is estimated at the considerable sum of 600 billion tons. This, without counting the huge mass of carbon contained in carbon, natural gas, and petroleum deposits, and also accumulated by living organisms.

Our singular composition betrays the fact that the chemical processes of living organisms are clearly distinct from those that occur in the inert matter of the planet's surface.

Why are these elements of hydrogen, oxygen, carbon, and nitrogen so abundant in living things? We find the answer in the ability of these elements' atoms to form covalent bonds.

As we have already seen, hydrogen can form one covalent bond, oxygen two, nitrogen three, and carbon four. Since these four elements can easily react with each other, an extremely wide range of possible combinations exists that determines a large number of different compounds. This variety of combinations is increased by the fact that carbon, oxygen, and nitrogen can share single or double bonds with other elements. Moreover, the compounds formed in this way are very stable, since covalent bonds among these four elements are found to be among the strongest and most stable.

Since carbon is the element that can form a greater number of bonds, up to four, it is the fundamental piece in this molecular *puzzle.* The union of carbon atoms can form linear chains, networks, closed chains (*rings*), or any combination of those structures, producing molecules of substantial size and complexity; and they will still have bonds free to unite with hydrogen, oxygen, and/or nitrogen atoms. Basically, all the molecules of living things are formed from a skeleton of bonded carbon atoms. In fact, carbon is the essential element in the chemistry of living things, and it is not surprising that life is often identified with carbon.

However, another element exists that is disposed to competing with carbon. This is carbon's column mate on the periodic table: silicon. Like carbon, silicon's electrical structure also enables it to form up to four covalent bonds. This fact has led some to propose that life would also be possible based on silicon's chemical composition, the same as with carbon. Some have even reached the conclusion that there could be living organisms on other planets composed of silicon instead of carbon. But if this opinion is correct, how do we explain that life on our planet originated from carbon and not from silicon, which is much more abundant? The fact is that silicon is not as good as carbon for forming living organisms owing to its different chemical properties.

In the first place, silicon has less ability than carbon for forming strong covalent bonds with itself. That is to say, the covalent bonds among the silicon atoms are less stable than those that are formed among carbon atoms. This situation greatly limits the ability of silicon to form chains, networks, and rings. Moreover, the union between silicon and oxygen, in the form of *silica* or *silicates,* is not soluble in water nor does it give off any type of gas. Silica and silicates form rocks, which would make it difficult for them to gain access to some hypothetical living organisms made of silicon. On the other hand, the union of carbon and oxygen produces *carbon dioxide*, an atmospheric gas that is soluble in water, or *carbonates* that, even though they can hasten and lead to rock formations (in fact, there are entire mountains formed of carbonates in the form of limestone rock), can be dissolved by water. The presence of carbon in the air and water facilitates its entry into living things. Finally, the union between silicon and oxygen is much more stable than that of carbon and oxygen so that, even though they might easily accede to silica and silicates, silicon's hypothetical organisms would have a very hard time acquiring them, since they would be unable to remove them from the oxygen. For their part, the union of carbon and oxygen, in the form of carbon dioxide, can be broken up, providing a source of carbon for living things.

Nevertheless, not all living things are capable of breaking down the carbon dioxide molecule; this heroic chemical deed belongs solely to *autotrophic* organisms, which we will discuss later.

In short, as organisms, we are extraordinarily demanding when choosing the chemical elements that go into our composition. Practically the only elements we use are oxygen, hydrogen, carbon, some nitrogen, and bits of others (especially sulfur and phosphorous).

But if we are so demanding where atoms are concerned, our singularity is even greater when it comes to molecules. Molecules are atoms linked together in precise quantities and with very tangible chemical and physical properties. Every substance that exists is comprised of one or more specific kinds of molecules. The *ozone* molecule consists of three oxygen atoms, the water molecule with two hydrogen atoms and one oxygen atom, and the carbon dioxide molecule with two oxygen atoms and one carbon atom. As you can see, each molecule has its own *recipe*, or *atomic formula*, which expresses how many atoms of each element comprise its composition. Thus, the respective formulas for ozone, water, and carbon dioxide are: O_3(three atoms of oxygen), H_2O (two atoms of hydrogen and one oxygen), and CO_2 (one atom of carbon and two oxygen).

All the different types of rock formations are comprised of a relatively small number of different kinds of molecules. In volcanic eruptions, these molecules can come together in ways that produce the molecules of most of the atmospheric gases, with the extraordinary exception of oxygen, which we will also discuss later. For their part, the salts dissolved in water also derive from the molecules in rock formations and can initiate new formations. What about the molecules in living things? Are they also the same type of molecules as those found in rock formations? No. We are composed of a group of molecules that are exclusive to living things.

However, if the molecules that constitute living organisms are found only in the organisms themselves, where did the first living being get its particular molecules? In order to answer that question (which is simply a chemical version of which came first, the chicken or egg?), we must first know, in broad outline, what are those very special molecules that constitute our formation.

Basically, the molecules of life, or *biomolecules*, can be grouped into four large classes: *glycosides* (or sugars), *lipids*, *proteins*, and *nucleates*. Moreover, even though it is not exclusive to living things, we should also devote a few lines to water; when all is said and done, it represents between 70 percent and 90 percent of the chemical composition of organisms.

Glycosides comprise a family of chemically homogeneous molecules. When their chemical composition was formed, *carbohydrates* were dominant because for each carbon atom, there *are* two hydrogen and one oxygen, as if each carbon atom was bonded to a molecule of water. In fact, this is not so, but the term carbohydrate has persisted and is more common than that of glycosides. Of the many kinds of existing glycosides, we will concern ourselves with the most popular one, which is also the most important one for living things: *glucose*.

The glucose molecule is comprised of six carbon atoms, twelve hydrogen, and six oxygen ($C_6H_{12}O_6$). As living organisms, we use glucose for two principal purposes: as fuel for energy and construction material.

Glucose molecules embody a large amount of chemical energy. For example, one teaspoonful of glucose contains enough energy to raise the temperature of one liter of water by almost 15 °C. Most of our cells can not only derive their energy from glucose, as we will soon see, they can also use lipids and proteins for fuel. But there is one exception to the norm: the nervous system. Under normal conditions, our brain cells can only use glucose as an energy source. Because of that, it is essential to maintain a constant level of glucose in the blood (about 0.8 grams of glucose for every liter of blood). If the amount of glucose in the blood drops to half that level, cerebral dysfunction appears, and if it is reduced to a fourth part, it can produce a coma. Our brain is a tireless devourer of glucose and requires around 140 daily grams of this glycoside, an amount that represents the energy required to raise the temperature of 533 liters of water by one degree centigrade!

On the other hand, glucose molecules can easily attach themselves to each other through covalent bonding. Every glucose molecule can bond with others at various points, thereby producing different types of linear chains and/or three-dimensional networks, each one with its own idiosyncrasies. These *super molecules*, formed by hundreds or thousands of glucose molecules, are generically called *polysaccharides*.

Some polysaccharides serve to store glucose, in the hope of

using it as fuel, while others are used as construction material for certain parts of living things. Fungi and living organisms store glucose in the form of a polysaccharide called *glucogenesis,* which in our case is specifically stored in our liver and muscles. The mission of the deposits of glucogenesis in our organism is to ensure the supply of glucose to the organs, especially the brain, under stressful situations, such as strenuous exercise. Under such circumstances, the supply of glucose is discharged on behalf of our body's cells, with the risk of the glucose level in our blood being lowered, and our cerebral function being affected. In these cases, the bonds between the units of glucose (which comprise the glucogenesis molecules) come apart, and the latter are released into the blood. When the stressful situation ends, the excess of glucose withdraws from the blood, with glucogenesis again being formed.

In plants, the principal reserve polysaccharide is *starch*. Because plant life is quieter than that of animals and does not have a brain that requires a constant supply of blood, starch does not have the mission of maintaining a stable level of glucose in plant tissue. Plants accumulate starch in order to face other situations.

For example, many plants take advantage of benign seasons in order to manufacture the greatest possible quantity of glucose, which they store in their subterranean organs (in fact, plants manufacture their own glucose, and later we will see how they do it). When winter arrives, these plants lose their stems and leaves, but their subterranean parts remain alive, subsisting thanks to the glucose that was stored during good weather. Potatoes are a good example of this kind of plant. At other times, plants accumulate starch in their seeds. They thus assure themselves that their embryos have a sufficient supply of glucose until they can acquire it. Cereals are this kind of plant.

Most of the glucose that reaches our organism comes from the starch in plants through potatoes, rice, and cereals, such as wheat (whose flour, besides going into making bread, is also used to produce nutritious pastas).

But plants use much of their glucose to build another type of polysaccharide: *cellulose*, the main component in roots, stems, and leaves. We humans use this substance to manufacture paper, cardboard, etc. Even though cellulose has the same chemical composition as glucogenesis and starch, the way in which glucose molecules are bonded together is different in each case. This difference determines their special physical and chemical properties. Thus, cellulose is a very consistent substance and also a wonderful thermal insulator. These characteristics are very useful for plant cells, which use cellulose to construct a wall that completely surrounds them, helping them to maintain their shape and insulating them against the elements.

Although the most notable difference between cellulose, on the one hand, and glucogenesis and starch on the other, lies in their biological nature: we cannot digest cellulose.

Perhaps you may be thinking that there is something here that doesn't add up. Since there are so many animals that exclusively feed off roots, stems, leaves, and even bark, at the same time, it does not seem possible that these plant organs are primarily comprised of cellulose, and animals and we humans cannot digest it. Still, that is how it is. What happens is that animals have consistent help from a series of microorganisms that live in huge numbers in their digestive tracts and break down the cellulose into glucose. Thus, on ingesting plant matter, animals are, in fact, nourishing the microbes in their digestive apparatus, and the microbes proliferate thanks to the supply of glucose that they obtain from the cellulose. Animal life is limited to ingesting microorganisms, together with the glucose that the microorganisms have acquired.

Naturally, enough microbes always remain in their digestive tracts in order to take care of their next meal.

Although we cannot digest it, nor do we depend on an adequate supply of microorganisms, cellulose plays a very important role in our nutrition. It has to do with an element called *vegetable fiber*, something that facilitates the work performed by our intestines, and the lack of it in our diet can cause serious disorders.

The second type of biomolecules is lipids, among which are *fats* and *oils*, and they are responsible for our shape. It is very difficult to chemically define lipids, since they constitute a wide-ranging class of molecules, although they all share one physical property: their insolubility in water.

Lipids fulfill discharge functions in organisms, comprising a very important source of energy; for example, cardiac tissue acquires most of the energy necessary for making the heart beat from lipids (70 percent, at normal rate, or shortly after increasing its activity). On the other hand, lipids are the main component in cellular membranes, and in one heterogeneous group of lipids, we find certain *hormones* (such as *androgen*, male sexual hormones, and *estrogen*, female sexual hormones) and *vitamins* (such as A, E, and K).

Delving deeper into the great chemical disparity in lipids and their biological polyvalent exceeds the purpose of this book, but there is a special class of lipids worthy of a few explanatory notes. These are *fatty acids*. Even though they are rarely found openly in cells, they are one of the fundamental components in many types of lipids. In accordance with their chemical structure, there are two kinds of fatty acids: *saturated* and *unsaturated* fatty acids. The difference is that carbon atoms in the first group are bonded together through simple covalent bonds, while carbon atoms in the second group are linked through double covalent bonds. If there is only one double bond, we are talking about a *monounsaturated* fatty acid, and if it presents two or more double bonds, it is called *polyunsaturated*. We humans produce rich fat in saturated fatty acids, while vegetable fats are more prodigious in unsaturated fatty acids. Animals that live in low temperature zones, and in whose fats there is also a predominance of unsaturated fatty acids, avoid this norm.

As mammals, we can synthesize practically all the saturated and polyunsaturated fatty acids that we need for life from simpler molecules ingested into our diet, but we are not able to generate polyunsaturated fatty acids, and so we must incorporate them with our food. If a laboratory animal is deprived of polyunsaturated fatty acids, it suffers numerous pathologies and ends up dying. The special importance of these fatty acids, which we must ingest, justifies their nomenclature as *essential* fatty acids; these are *linolenic* acid and *gamma-linolenic* acid. Fortunately for us, these essential fatty acids are very abundant among plants.

This situation could lead us to think that a diet completely lacking in plant foods is not possible. However, Eskimos have seen to it to prove the contrary: there are no plants to which they can turn in their ecosystems, and therefore, the vegetable component is absent from their diet (at least, until a few years ago). Where do Eskimos get their essential fatty acids? The answer lies in the once repudiated and now accepted bluefish. This type of vertebrate is able to synthesize its own polyunsaturated fatty acids, and Eskimos eat this type of fish or other animals that are also predators of these fish.

Moreover, fatty acids are also important for our health because of their relationship with a molecule whose name alone makes us shudder: *cholesterol*, a lipid that, paradoxically, is indispensable to

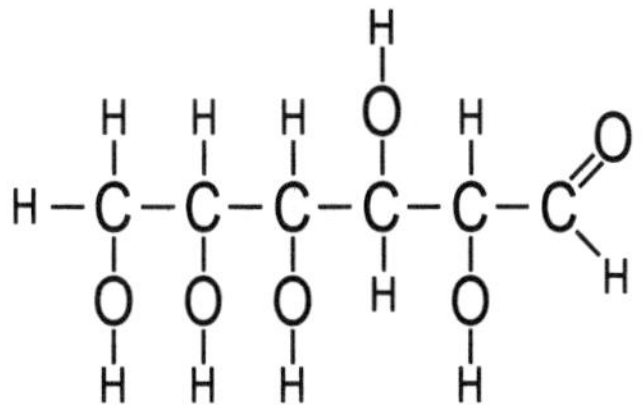

Glucose, monosaccharide

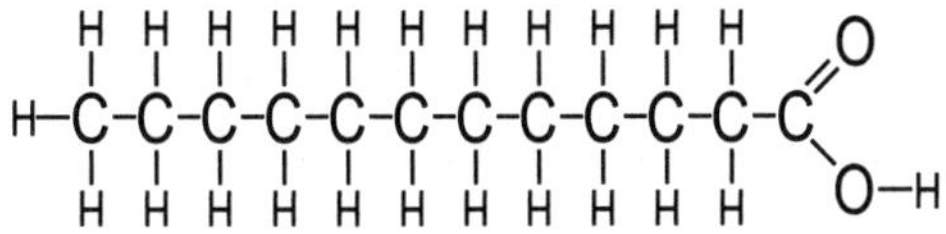

Lauric acid, saturated fatty acid

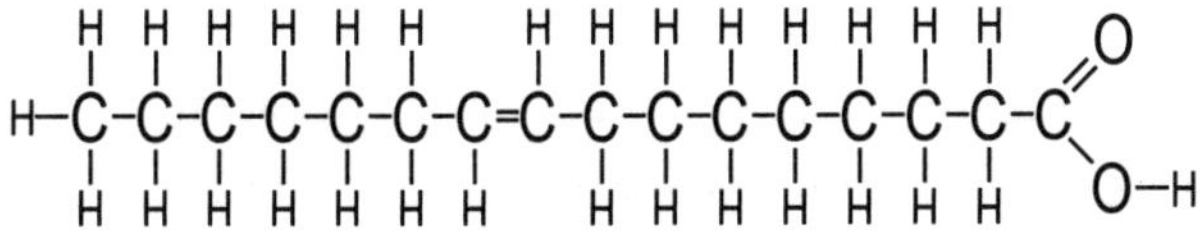

Palmitoleic acid, monounsaturated fatty acid

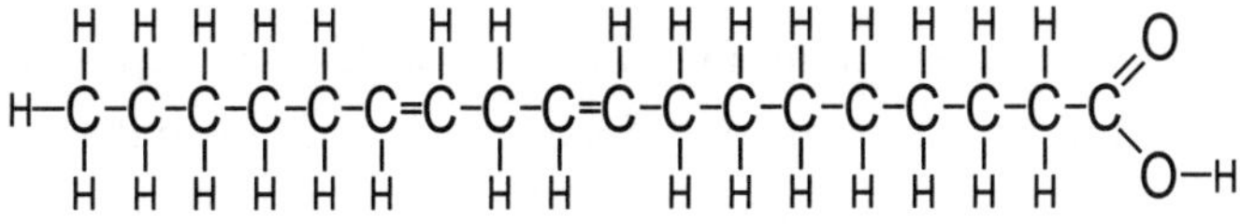

Linoleic acid, poliunsaturated fatty acid

Figure 7: *A schematic illustration of glucose and different fatty acids. It can be proven that hydrogen always forms a covalent bond, two oxygen and four carbon.*

the correct functioning of our organism. Among its other missions, cholesterol is a fundamental component in the membrane of animal cells and is a precursor, among other molecules, of *biliary salts* and *steroid* hormones (androgen and estrogen). However, and despite its great importance, cholesterol can become very dangerous to our health. When the cholesterol level in our blood exceeds a certain limit, it tends to become deposited in the internal wall of the blood vessels, collaborating in the formation of *atheromas*, which can begin to impede the flow of blood to the heart and brain.

It has been proven that a direct relationship exists between the cholesterol level in the blood and the amount of cholesterol and saturated fatty acids ingested in one's diet. In other words, foods rich in animal fats, which contain cholesterol and saturated fatty acids, can cause an increase in the cholesterol level in the blood if they are not taken in moderation. On the other hand, it is known that increasing the proportion of polyunsaturated fatty acids in one's diet causes the cholesterol level in the blood to decrease appreciably. These findings have vindicated the so-called Mediterranean diet, superb in vegetable fats and bluefish rich in polyunsaturated fatty acids.

Proteins are the third group of biomolecules, and together with nucleic acids, the most noteworthy. This assertion is based on several facts. In the first place, the most abundant biomolecules in cells constitute about half of their dry weight. In addition to many other functions, proteins take part in all the chemical reactions in living things, even to the point of being responsible for their characteristics and functioning. Proteins hold such great importance for organisms that the structure of each one is codified in the cells' genetic material. This means that the biological information, which is transmitted from generation to generation, basically consists of *recipes* for synthesizing the right proteins.

We are dealing with enormous molecules, constituted, in turn, from the union of other smaller molecules called *amino acids*. In the chemical structure of any amino acid, two portions can be identified: one that is uniform in all of them, and the other that is variable

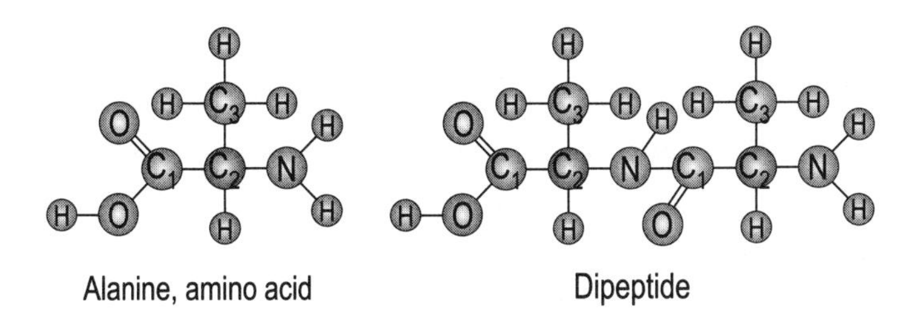

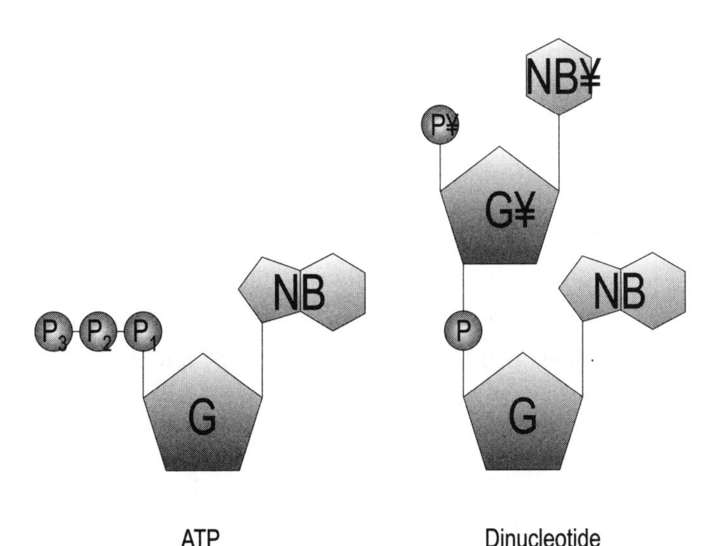

Figure 8: *Upper left: A schematic illustration of an amino acid (alinine). The carbon atoms corresponding to the acid group and variable region are named C1 and C3, respectively. The letter N identifies the nitrogen atom in the amino group. Upper right: Two amino acids covalently bonded (dipeptide) by means of one acid group (C1) and the amino group (N) of the other. Lower left: An illustration of an adenosine triphosphate molecule, or ATP. The letter G indicates the glucide of five carbon atoms, NB identifies the nitrogenous base (adenine), and the three molecules of phosphoric acid are labeled P1, P2, and P3. The bonds between these molecules of phosphoric acid are high-energy bonds in which the cell stores energy. Lower right: Two nucleotides bonded through a molecule of phosphoric acid (P), the same as in the nucleic acids (DNA and RNA).*

and determines the chemical properties of each type of amino acid. The uniform portion includes an *amino* group (one nitrogen atom bonded to two hydrogen atoms through covalent bonds) and an *acid* group (one carbon atom bonded to two oxygen atoms: one of them through a double covalent bond, and the other is bonded to one hydrogen atom through a single covalent bond). The amino group of one amino acid can react with the acid group of another amino acid and retain its covalent bond. In this way, amino acids can link up like train cars to form chains of quite variable lengths. Most proteins are composed of anywhere from 100 to 300 amino acids, but this number can be much greater, reaching about 2,000 amino acids in some proteins. Surprisingly, the number of different kinds of amino acids is greatly reduced in living organisms. The vast majority of the tens of thousands of proteins in the various organisms are combinations of only twenty different types of amino acids.

We vertebrates, however, are not able to synthesize every one of those twenty types of amino acids. So, we humans need to incorporate ten, which are designated as *essential amino acids,* into our food.[3] Of these, there are five types that are relatively rare in cereals and leguminous plants, which means that people who decide to maintain a strictly vegetarian diet must plan this with utmost care to assure their bodies the proper supply of essential amino acids.

Proteins fulfill an astonishing variety of functions in living organisms. Even though it is not their principal mission, cells also use them as one of their common fuels. A consequence of the degradability of proteins is the production of a nitrogenous residue (originating from the amino group of amino acids) that

3. Throughout our lifetime, the number of amino acids we need to incorporate into our diet diminishes. This is because we are able to independently synthesize, though in small quantities, some of the amino acids labeled essential. During growth, the organism needs a large supply of amino acids to build new tissues. Under those conditions, our ability to synthesize those amino acids is more than sufficient, and we must get them from our food.

must be eliminated, basically, in the urine (in the form of *urea* in land-dwelling mammals). This process allows an individual to eliminate between 6 and 20 grams of nitrogen daily.

Another of the functions carried out by proteins is to serve as construction material for different tissues and structures. Cosmetic ads have familiarized us with some of the proteins engaged in this work, such as *keratin* and *collagen*. The first is the fundamental component in nails, hair, and the skin's outer layer. Collagen is found, above all, in those tissues that require resistance and elasticity at the same time: the skin, tendons, ligaments, or the walls of blood cells, for example.

Proteins are also involved in the body's defense against microorganisms (*pathogens/immunoglobins*), in the transport of numerous substances, like oxygen (*hemoglobin*) in muscular contraction (*actin* and *myosin*) and in hormonal function (*insulin* and *growth hormones*, for example).

But the most characteristic mission of proteins is to serve as *biocatalysts* in living things. A catalyst is a substance that increases the speed of a chemical reaction without participating in it. The catalyst in automobile engines accelerates gasoline's combustion without mixing with the oxygen or the gasoline. However, nearly all the different chemical reactions that take place in living organisms would not occur, or would do it too slowly, if they were not catalyzed by a family of proteins known as *enzymes*. It can be said that enzymes regulate and direct the chemical reactions of living organisms. Therefore, we must not be surprised that biological inheritance consists of transmitting the necessary information to descendents in order to synthesize the proteins that characterize each species. Nucleic acids head up this mission, the conveyance of biological inheritance.

Glucides and lipids are composed of carbon, oxygen, and

hydrogen. Nitrogen (and a little sulfur) is added to these elements in the chemical composition of proteins. Nucleic acids add a fifth element (besides carbon, oxygen, hydrogen, and nitrogen) to their chemical composition: phosphorous. Just as with proteins, nucleic acids are composed of the union of other lesser molecules: *nucleotides* in this case.

Nucleotides are very complex molecules formed, in turn, by the bonding of three other simpler molecules: *phosphoric acid*, a *nitrogenous base*, and a glucide made of five carbon atoms. Glucide (or sugar) constitutes the cornerstone in this chemical trio: the nitrogenous base and phosphoric acid always join it through covalent bonds. In turn, the molecule of phosphoric acid couples with the sugar of another nucleotide, whose molecule of phosphoric acid attaches itself to the sugar of a third nucleotide, and so on, in succession, in order to form chains of several hundred to several hundred million nucleotides, according to the type of nucleic acid.

All nucleotides display the same type of phosphoric acid (*orthophosphoric acid*), but the same thing does not happen with the sugar molecule or with nitrogenous bases. In fact, there are two types of sugars that determine two different kinds of nucleic acid and five different nitrogenous bases. Both sugars are *ribose* (in fact, the D-ribose) and its chemical relative *deoxyribose*, similar to ribose but with one less oxygen atom.

The nucleotides formed with ribose can be bonded with four types of nitrogenous bases: *adenine, guanine, cytosine, and uracil.* These nucleotides constitute a type of nucleic acid called *RNA* (the initials for its name: *ribonucleic acid*). For their part, nucleotides constructed with deoxyribose can bond with the same nitrogenous bases as those composed of ribose, except for uracil, whose place is taken by the fifth type of nitrogenous base: *thymine*. These nucleotides are the ones that compose *NDA* (*nucleic deoxyribose acid*). These chemical differences are extremely important because they make sure that the NDA and RNA carry out very different missions in the transmission of genetic information.

The different nucleotides, but especially those that include deoxyribose and adenine, can form other molecules, different from the nucleic acids and with a quite distinct mission. When they include three phosphoric acid molecules in their structure, instead of just one, nucleotides bond with each other to form chains, except that they remain isolated. These nucleotides with three phosphoric acids constitute the units of energy stored in cells; the most widely used these is *ATP* or *adenosine triphosphate.* We will also encounter this molecule later and will have an opportunity to understand its important role in cellular life.

One final note in this presentation on nucleic acids. *All* living things on the planet use the same types of nucleic acids, comprised of the same molecules, in accordance with the same chemical reactions. This is not by chance; it is a matter of the most direct evidence of the common origin of all living things, obvious proof of evolution.

It remains a paradox that we have given to our world, which was baptized with the name Earth, the nickname of blue planet, in reference not to the earth, but to the component that covers most of its surface: water. Water is an authentically singular substance, as much for its physical properties as for its chemical characteristics. Since its distinguishing features determine in great part the properties of living things, it is worth our while, although it has nothing to do with biomolecules, to devote the last lines of this chapter to water.

As we have already said in the section dedicated to the covalent bond, one molecule of water is created by one atom of oxygen forming a covalent bond with two hydrogen atoms. We also explained that the oxygen atom has a greater affinity for electrons that participate in the molecule's two bonds than the respective hydrogen atoms. As a consequence, these electrons are situated closer to the oxygen atom than the hydrogen atoms, causing a small negative electrical charge to appear next to the oxygen atom, as well as an equally small positive charge beside each hydrogen atom. Thus, water is a bipolar molecule; that is to say, one part positive and one part negative.

Life is based on a set of chemical reactions in which the different compounds involved either have to come together or separate in order to produce a reaction. For that to happen, they must be located inside a liquid environment. Very few chemical reactions can be produced in a solid since atoms must be able move around in order to react with each other. Atoms freely move inside a gas, but their density is so low that it diminishes the probability of their being able to react to each other.

On the other hand, liquids are ideal as a *facilitator* for chemical reactions. In liquids, atoms can freely move, and their density can be substantially increased, although for that to happen, they must dissolve in the liquid. So then, the ideal medium for chemical reactions is a liquid with a solvent ability. However, if the liquid is a very strong solvent, it can break the covalent bonds, impeding the formation of many of the biomolecules that are indispensable for living things.

The bipolar nature of water transforms it into a wonderful solvent. Let's imagine a table salt molecule, formed by an ionic bond between one atom of sodium and one atom of chlorine. Perhaps it is well to remember at this point that the ionic bond is one that is created by the existing electrical attraction between a positively charged atom, a result of its having lost one electron, and another atom, negatively charged, a result of its having accepted one additional electron. In table salt (sodium chloride), the sodium is the atom with the positive charge, and the chloride is the atom with the negative charge. Both remain bonded by the force of the electrical attraction. However, if we include a molecule of sodium chloride in water, the water molecules will position themselves so that the ones closest to the sodium atom will offer it their negative side (that of the oxygen atom), while the water molecules in proximity to the chloride atom will rotate in order to offer it their positive side (that of the two hydrogen atoms). In this way, a good number of water molecules will attempt to attract the sodium atom, while others will do likewise with the chloride atom. The result will be the fragmentation of the sodium chloride atom's ionic bond, which will separate into sodium and chloride atoms that will dissolve in the water. The sodium and chloride atoms can remain separated (dissolved) in the water because their electrical charges will be counterbalanced by the combined action of the weak charges of the water molecules that surround them.

The ability of water to dissolve ionic bonds, as well as to stabilize atoms (or molecules) with electrical charges, is fundamental in the chemistry of living things and makes water the ideal medium for such reactions. Moreover, water is not able to dissolve covalent bonds, thus allowing different biomolecules to form and remain stable.

The bipolar nature of water molecules is also responsible for the singular physical properties of this liquid. The hydrogen atoms in each water molecule, with their weak positive charge, are attracted by the slight negative charge of the oxygen atoms of other water molecules and are coupled with them through weak electrical bonds. These weak bonds are continually forming and separating, and as a result, even though the water molecules can freely flow back and forth among themselves, they are, at the same time, also linked to each other through a network of weak hydrogen-oxygen electrical bonds.

As a consequence of the existence of this network of weak bonds, it is difficult to increase the movement of the water molecules when the water is heated. Let's remember that the body's temperature is nothing more than the aggregate movement of its molecules. However, when water is heated, the energy that is generated primarily goes into breaking the electrical bonds among the water molecules, and but one part of that energy can be used to increase the molecules' movement. As a result, water can absorb a great deal of heat, its temperature increasing only slightly as it does so. This phenomenon is known as *thermal inertia* and occurs inversely as well: water can relinquish a lot of heat, its temperature only slightly diminishing in the process. In short, water heats up and cools down very slowly.

This characteristic of water explains why coastal regions enjoy milder climates than those that are inland. Seawater (and also the water in the atmosphere, which determines humidity) can absorb a lot of heat from the air in the summer, to give it back again, slowly, during the winter. Thus, summers are less stifling, and winters milder. In fact, this *thermostatic effect* of the ocean goes to work in the planetary sphere once the water of the warm regions, infused with heat, moves in the form of oceanic currents toward the colder zones, where it releases the aforementioned heat (there are also cold currents that carry water at low temperatures to warm regions). In this way, water distributes solar heat throughout the planet, creating conditions favorable to life.

An illustration of this is the water in the Gulf of Mexico, which absorbs an enormous amount of heat. This warm water runs, in the so-called Gulf Current, through the Atlantic, eastward until it reaches the western European coasts. The effect of this warm current on the European coasts is easy to check if we take into account that Paris, a city of temperate climate, rests on the same latitude as the icy coast of the Labrador Peninsula in Canada (along whose coasts runs a cold current that comes out of Greenland).

Another physical property of water of great interest is the fact that its boiling and freezing points are separated by 100°C (212°F)—a big interval. Since a large part of Earth's surface displays a temperature ranging between those two points throughout the year, most of the planet's water remains in a liquid state, ideal for life.

Even in regions where the temperature drops below freezing for a few months, water remains in a liquid state for a good part of the year. This is the result of another fortunate characteristic of the liquid element: its solid state (ice) is less dense than its liquid state. As a consequence, ice floats on water. This situation is unusual, given that in other compounds, it is normal for the opposite to happen: the solid phase is denser than the liquid phase and sinks.

Let's imagine what would happen in large areas of the ocean and in many lakes located at high altitudes if water were to behave like most compounds. During winter, water on the surface would freeze over as it surrenders its heat to the cold air. On becoming denser, this frozen layer would sink to the bottom, leaving a new layer of water in contact with the cold air. The layer of water on the surface would freeze again and once more sink to the bottom. The lake, or the sea, would gradually freeze from the bottom up, and a great mass of ice would form on the bottom. With the arrival of warm weather, the sun's rays would only manage to heat the uppermost region of the water, thereby leaving the layer of ice on the bottom unchanged. The following winter, new layers of ice, which could not be melted during the summer, would be added.

Finally, a winter would come along in which the lakes and oceans of the higher latitudes would be completely frozen from the bottom up. During the summer, only the uppermost layer, which would absorb the sun's heat, could melt, impeding the deepest part of the waters from thawing.

Fortunately for life at higher latitudes, the very opposite happens. During the winter season, the upper layer freezes, but because ice is less dense than water in its liquid state, it floats on the surface and remains there, insulating the deepest part of the water from the cold air and keeping it from freezing. When summer arrives, the sun's heat melts the external layer of ice. With the arrival of winter, a new cycle of freezing and thawing will begin, affecting only the water's upper layer.

In short, the molecular structure of water confers on it the ideal characteristics for embracing life: it is a weak solvent that contributes to the distribution of solar heat throughout the planet and remains in a liquid state in most regions of our world for a good part of the year. So great is the importance of water's properties for living things that it, together with its great abundance on Earth's surface, would justify our changing the name of our planet, as the song says, from Earth to Water.

Chapter IV
The Origin of Life

Knowledge of Nature

As living beings, we come to the world endowed with a sweeping knowledge of the natural sciences. This wisdom is nearly encyclopedic among animals that we see around us in our daily lives: birds and mammals. Stop to observe the sparrows where you live, or dogs and cats (whether your own pets or those of others); their behavior is very complex and wonderfully adjusted to the environment in which they live. They know exactly what things they should eat and not eat. They can recognize the members of their own species from among the many animals. They also distinguish, among their group, between male and female, between adults and those that have not reached maturity. They recognize the ideal season for mating, and what they must do to carry it out. Many animals can migrate to very distant locations, and they know the route to be followed, as well as when to undertake their journey. There are birds that display great knowledge of architecture in the construction of their nests, and spiders could give courses on the art of weaving.

What is certain is that entire books could be written, dedicated to the astonishing knowledge that animals have of their surroundings (in fact, they have been written and continue to be written every day), and we will take a look at this later. We humans are so convinced of this knowledge that it is a generally accepted argument that animals are capable of knowing, before we do, when a natural catastrophe is going to occur.

In the case of plant life, such knowledge is not as evident. Plants do not seem to present an example of astuteness, but still, they also know whether it is day or night, recognize the passing of the seasons, and act accordingly, blossoming with leaves, flowers, and fruits, or dropping them. They can also distinguish the direction of the light that reaches them or their spatial position (place a plant in a reverse position, and in a short time, you'll see that the stem has turned in order to resume its normal position). Their roots recognize terrain and favor those regions richest in nutrients or water. Some are even able to detect the presence of insects and trap them. They also possess certain attributes to attract those insects that will carry their pollen to other plants of their species. In short, they too display a great knowledge of nature. No doubt, it is to a lesser degree than in the case of most animals, but the fact is, that in order to survive, a plant does not need to know as much about life as an animal.

It might seem that microbes cannot be included among living things that possess *knowledge*, but this is not so. They also distinguish between what is food and what is not food, or perfectly recognize what area of their surroundings is more favorable to life. Moreover, many microorganisms are parasites of multicellular organisms, but they do not attack all the cells of their unfortunate host, only certain types. The sickness known as malaria is one of the principle causes of mortality (alone or in association with other pathologies) in third-world countries. The agent responsible for this terrible illness is a microorganism belonging to the genus

Plasmodium (in fact, there are several species of *Plasmodium* that produce different strains of the sickness; the most lethal of them all being *Plasmodium falciparum*). However, this microbe is extraordinarily inventive and only infects the red blood cells in our blood. In other words, it is capable of recognizing a specific cellular type from among the many that make up our body; doubtless, it is a microbe with large elements of cytology.

We humans also have a great knowledge of the natural world that is different from the knowledge all other living beings possess, both quantitatively and qualitatively. In the first place, our knowledge is substantially greater than that of any other organism, or than that of all of them put together. On the other hand, the knowledge that all the other organisms posses is, for the most part, innate, while the overwhelming majority of our knowledge is acquired, through experience or learning throughout our lifetime. In other words, all other living organisms are already born with almost everything (or its totality) they need to know, much of this knowledge being codified in their genes in the form of common behavior patterns (*instincts*, if we are talking about animals). But in the case of humans, this is not so. We come to the world with only a small part of the knowledge that we acquire during our lives. The main body of our knowledge is not inscribed in our genes, but in the minds of those who preceded us, and more recently, in our books.

There is another fundamental difference between our knowledge of nature and that of other organisms. While theirs lacks an understanding of natural phenomena, ours is based on a comprehension of it. As a consequence, all other living organisms passively behave in the medium of nature, not having the ability (or the greatly reduced ability) to modify the conditions of their environment in order to adapt them to their needs. On the other hand, our knowledge of the mechanisms that govern nature allows us to deal with them and greatly modify our surroundings, with the benefits and dangers that ensue.

These two very human qualities, the faculty for comprehending the laws of the natural world and the ability to manipulate them, are separately designated as science and technology. But we have not always made uniform use of them. For most of history (including pre-history), man believed that the explanation of natural phenomena belonged to the supernatural.

Different cultures have coincided in embellishing many diverse myths in order to explain natural phenomena: birth, death, rain, lightning, the origin of the world, of life . . . And since, according to that point of view, the causes for the events of this world lie outside it, different civilizations have devised endless ways to approach the supernatural world and influence it in the way they wished, through rituals, spells, sacrifices, incantations . . .

But there was a moment and place in which a radically different thought emerged. Perhaps it was the most important intellectual revolution in the history of mankind. The place was Greece, and the time was around the seventh century B.C. The new idea that was born, and that forever changed the course of history, was this: *natural phenomena have natural causes*. But the ancient Greeks went even further, claiming that man could understand those natural causes. Both affirmations contradicted what all cultures had believed since the dawn of time, and they are the basis for our modern conception of nature, marking the birth of science and technology such as we understand them today.

Ever since the baroque era, this concept of a self-sufficient universe, a universe not needing supernatural forces to explain it, gradually gained ground every time the greater reality of matter's elemental properties was recognized. At the beginning of the nineteenth century, it was the phenomenon of life itself that defied explanation from this perspective, and thus it constituted the last stronghold of those who defended the presence of supernatural forces in the natural world.

The great development of biology throughout the nineteenth and twentieth centuries allowed us to assert that the manner in which living things operate is governed by natural laws like all other matter comprising the universe. Without a doubt, one of the most transcendental milestones in our understanding of the biological phenomenon was the unequivocal proof of the process of evolution and the proposition of a natural mechanism that explained it. Afterward, just two points of contention remained among those who thought life to be a natural phenomenon and those who stood in support of its supernatural origin. Both questions correspond to two unique phenomena and are difficult to explain from the camp of those who were seeking natural explanations: the origin of life and the dawn of *consciousness* (or man, if you prefer). It is still too soon, in this chapter, to concern ourselves with the second of them, but this is the right moment to focus our attention on the first—the origin of life.

A Bit of Bad News

The classic concept of modern science is based on the conviction that nature must display *regularities* in its manifestations. Given that laws that emanate from the basic properties of matter govern the universe, it then follows that every time the same conditions appear, the same effects must ensue. This kind of scientific premise is known as *determinism*: the result of a process is predetermined by the conditions at the outset. From this perspective, the scientific task consists of finding and understanding these regularities, while technological activity pursues the control of the corresponding conditions in order to produce the desired effects.

However, since the end of the nineteenth century, we have known that there are many natural processes that can, assuming the same conditions, yield multiple results, some being more probable than others. This is a vision of the natural world as a set of *probabilities*: the result of a process is not rigidly determined by the initial conditions, unless possible multiple results exist that have different probabilities of being realized. In the concept of probabilities, it is accepted that *peculiarities* are produced in nature—events that, because of their low probability of occurrence, are not likely to be repeated.

From this perspective, we can ask ourselves if the manifestation of life from inert matter is a regular or singular phenomenon in nature.

For Aristotle, the question left no room for doubt. Life naturally originated from inert matter. In his works, he described cases of insects, crustaceans, mollusks, fish, and mice appearing as originations from clay. In giving a name to this process, by which life originated from inert matter, Aristotle coined the expression *spontaneous generation*. Aristotle's "spontaneous generation" was a natural cause of the phenomenon of life and a regular manifestation. Even though supporting a very different conception of nature from Aristotle's, Democritus and Epicurus also believed life to be a phenomenon that spontaneously originated from inert matter. Their disciple, Titus Lucretius Carus, expressed it this way in the following verses:

> We may at last see the lowly worm
> Born out of the repulsive mud,
> When the wet Earth has been putrefied
> By abundant rain [. . .]

It was Aristotle's great intellectual authority that contributed to the concept of spontaneous generation being accepted without question until the middle of the seventeenth century. In fact, it was supported by some of the great thinkers of the Church, like Saint Basil (329–379), Saint Augustine, and Saint Thomas of Aquinas. But it was not only Aristotle's prestige, and that of the saints, which gave credence to the theory of spontaneous generation. It was also the matter of a phenomenon that was apparently easy to prove experimentally. In their day, highly prestigious personalities in the world of science—such as the English physician William Harvey (1578–1657), the French surgeon Ambroise Paré (1510–1590), or the Belgian physician Jean-Baptiste Van Helmont (1577–1644)—all supported the existence of this phenomenon. Van Helmont even proposed a method for producing mice, in three weeks' time, by placing grains of wheat together with a soiled shirt in a vessel, causing a reaction that would transform the wheat into mice. Isaac Newton accepted spontaneous generation as a fact.

Nevertheless, the credibility of spontaneous generation began to collapse as a result of the experiments of Dr. Florencia Francesco Redi (1626–1697), a physician who took up the example of worms that *spontaneously* appear from decomposed organic matter. In his opinion, worms were the consequence of the insemination of organic matter by other living organisms, flies in particular, which deposited their eggs on it. In order to prove it, in 1668, Redi concocted a very simple experiment: he collected decomposed organic matter inside two vessels, then covered one of them with a muslin cloth and left the other one open. After a certain passage of time, worms appeared inside the open vessel, but not in the one covered with muslin. By covering the one vessel, Redi had prevented flies from getting at the decomposed matter and depositing their eggs on it.

Redi is normally considered the first scientist to discredit the theory of spontaneous generation through experimentation, but the fact is, upon failing in his attempt to explain the phenomenon of the *spontaneous* appearance of insects inside the excrescence—known as *galls* (tissue swellings)—of certain plants, Redi had no choice but to accept a special version of spontaneous generation in order to justify the phenomenon in question. It was the Italian physician and biologist Antonio Vallisnieri (1661–1730) who, around 1700, unequivocally proved that insects are hatched from galls after the adults of the same species have laid their eggs there. The plant reacted against the foreign body—the insect's egg—that had been introduced into it and generated a type of anomalous growth around it: galls.

Redi's experiment and Vallisnieri's work dealt a hard blow to the theory of spontaneous generation, which over time was submitted to ever increasingly rigorous observations on the *spontaneous* appearance of other invertebrates, which always ended up being explained as the result of preceding organisms having laid their eggs. In this way, the theory of spontaneous generation gradually fell into discredit.

However, the theory was resuscitated not long afterward because of the discoveries made by one of the great inventions of the seventeenth century: the microscope, which allows us to see what is invisible to the naked eye. One of the most famous assemblers of microscopes of that period was a Dutch tradesman, a contemporary of Redi: Antonie van Leeuwenhoek (1632–1723).

In addition to his skill in building microscopes, assembling more than three hundred in his lifetime, Leeuwenhoek made important scientific discoveries, such as the existence of microorganisms (in 1683), red blood cells, and sperm cells. Leeuwenhoek contributed to the loss of spontaneous generation's prestige when he demonstrated that organisms, like mussels, fleas, and weevils, do not spontaneously emerge from sand, but rather are the result of minute, preexisting eggs.

Along with Redi's and Vallisnieri's work, the evidence contributed by Leeuwenhoek caused the idea of spontaneous generation to be abandoned insofar as complex organisms were concerned, but paradoxically, the discovery of the existence of microorganisms offered unexpected support to the theory. No one thought that microbes could be reproduced through *seeds* or *eggs*, therefore their appearance in a medium in which they did not previously exist could only be explained by spontaneous generation. The proof of this phenomenon was outside of anyone's reach: it required boiling a liquid, to which organic matter had been added, in order to rid it of any microbes; then after allowing a few days to pass and turning the microscope on it, a thriving abundance of microorganisms was discovered. This was conclusive proof of spontaneous generation, according to its defenders. The phenomenon had remained restricted to the environment of microscopic life, but on the other hand, the evidence seemed irrefutable.

For Leeuwenhoek, these results were not conclusive, and he suspected that the microorganisms that showed up in the culture mediums had come about, after sterilization, from the air. In fact, a follower of Leeuwenhoek, Luis Joblot (1645–1723), carried out an experiment, in 1718, to a certain extent similar to the one performed by Redi, consistent with boiling a culture medium for fifteen minutes in two separate vessels, one open and one closed. After the culture medium was boiled, microorganisms quickly filled the open vessel, while the one that had been covered remained sterile for several days; but when the cover was removed, microbes rapidly appeared. Even though the outcome of the experiment seems conclusive regarding Leeuwenhoek's suspicions, most scientists of the day remained unconvinced. Many believed prolonged heating had altered the composition of the air inside a closed vessel. This alteration, it was argued, caused the air to lose its ability to generate life, preventing the spontaneous appearance of microorganisms.

Thus, the polemic continued throughout the eighteenth century. Among those opposed to spontaneous generation was the Italian priest and scientist Lazzaro Spallanzani (1729–1799), a versatile researcher among whose achievements are having performed the first head transplant in history . . . between snails, artificial insemination of a female dog, and the presentation of evidence that it is in the tissues of living organisms where oxygen is converted into carbon dioxide (and not in the lungs, as Lavoisier had erroneously suggested).

Spallanzani's adversary in this polemic was another Catholic priest, the Englishman John Turberville Needham (1713–1781). Needham was a staunch defender of spontaneous generation and completed a series of experiments, presenting their results as indisputable evidence of the phenomenon in question: he took a series of vessels and placed a lamb culture inside, boiling both vessels for two minutes, after which he covered them, and at the end of a certain period of time, he observed them, finding a great number of microorganisms. Needham's conclusion was that the microorganisms in question had appeared through spontaneous generation, given that the liquid in the containers had been sterilized and containers properly covered.

But Spallanzani did not concur with Needham on the interpretation of the evidence. In his opinion, Needham's treatment was too lax and did not guarantee the liquid had been sterilized. Moreover, on covering the vessels *after* the liquid had come to a boil, Needham had allowed the microorganisms from the air to contaminate the vessels once more. In order to prove his point, Spallanzani boiled the culture medium in the hermetically sealed vessels for forty-five minutes. Microorganisms did not appear in Spallanzani's vessels. Needham objected, saying that as in the case of Joblot's experiment, Spallanzani's treatment had altered the very nature of the air, thereby making it impossible for spontaneous generation to take place.

The dispute remained deadlocked in the impossibility of finding a method that would permit the vessels to be sterilized without their interior air being affected by the process. So it was that by the end of the eighteenth century and beginning of the nineteenth, great biologists like the Frenchmen Georges-Louis Leclerc de Buffon (1707–1788), Jean-Baptiste de Monet, Baron de Lamarck, and Georges Cuvier (1769–1832), or the Englishman (and Darwin's grandfather) Erasmus Darwin, remained partial to the theory of spontaneous generation. Paradoxically, among the personalities of the period who did not accept spontaneous generation was one who did not come from the field of biology; François-Marie Arouet, *Voltaire* (1694–1778), published a pamphlet in 1769 in which he satirized Needham's work and praised that of Spallanzani. But the question of spontaneous generation, and especially its importance with regard to our view of the world, was going to experience a decisive turn by the middle of the nineteenth century.

Until that moment, the debate over spontaneous generation was relatively free of any ideological encumbrance, the reason being that for most researchers and philosophers, the human being did not constitute part of nature. Or at least not in the same way as other living things. The prevailing thought was one that accepted the human species as being different from all other organisms, as having appeared because of a special act of divine creation. So then the idea that microbes, or even invertebrates, could or could not spontaneously appear, continued to be an academic dispute, separate from the question of the origin of man. Because of this dispute, it was possible for two Catholic priests, Spallanzani and Needham, to unequivocally align themselves with one camp or the other regarding spontaneous generation.

But this situation changed forever in 1859. That was the year Darwin published his work *The Origin of the Species*, in which he indisputably demonstrated the phenomenon of *evolution*, according to which all living organisms are linked through a series of common ancestry, and the farther back in time, the greater the morphological difference among them. Meaning this: we humans share a common ancestor, not so distant in time, with primates most like ourselves (chimpanzees, orangutans, and gorillas), an ancestor farther removed from all other mammals, and another, even more distant, from all other terrestrial vertebrates . . . and so on, until we arrive at a single common ancestor with the remaining organisms: the first living being.

In a single stroke, man became part of the animal kingdom, and the question of our origin remained linked to that of all other creatures. From this perspective, the old polemic of spontaneous generation acquired a completely different nuance. The problem of the origin of man was now made clear. For the defenders of the divine origin of man, only two alternatives were left: either Darwin was wrong and the origin of man was not linked to other living organisms (in which case, the question of spontaneous generation would continue being an *academic* subject), or if on the other hand Darwin was right, the origin of life (and therefore of man) must be linked to a single act of divine creation, incompatible with spontaneous generation.

The truth is that even though most of the scientific community quickly conformed to Darwin's ideas, the question of spontaneous generation remained a difficult problem to solve.

We have already established the technical difficulty involved in effecting an experiment that would satisfy both sides. Let's remember that the key to the problem, the Gordian knot of the debate, lay in acquiring a means for sterilizing the medium culture and air inside the vessel, without the air itself being subjected to the process of sterilization. Something that seemed impossible, given that since Joblot's experiment (in 1718), no one had managed to find the solution.

In an effort to break the deadlock in which the question was mired, the Paris Academy of Sciences offered, at the beginning of the decade, in 1860, a monetary prize to anyone who could settle the old debate. Little could the academicians imagine that the definitive solution was about to be reached at the hand of a French scientist who, in 1860, wrote to a friend: "I hope shortly to take a decisive step, emphatically resolving the celebrated question of spontaneous generation."

Who was this scientist who dared to announce the definitive solution to a debate that had lasted nearly two hundred years? It was one of the great minds in the history of science, Louis Pasteur (1822–1895). At the time he wrote those words, Pasteur was thirty-eight years old and already a renowned scientist for his discoveries in the chemistry of living organisms. At the age of twenty-six, he had presented, before the Paris Academy of Sciences, a work in which he explained the phenomenon of *isomerism,* or the spatial asymmetry of molecules, wherein two molecules with the same chemical composition and identity, but conjectural images of each other, display different physical and biological properties.

Explained in that way, the phenomenon of isomerism seems terribly complicated, but think about your hands: both have the same composition and structure, but each one is the theoretical reflection of the other (meaning that in a mirror, the right hand is like the reflection of the left hand), and this determines functional differences in both. This is true to the point that many objects of everyday use (the computer keyboard mouse, for example) are made to be specifically used by the right hand (if you are left-handed, in this world of right-handed people, you know what we are talking about). Clearly, hands are not molecules, but the analogy is useful in understanding the phenomenon of isomerism.

But let's return to Pasteur. In 1857, he proved that the process known as fermentation, of enormous industrial importance, was caused by the action of microorganisms (yeast and bacteria). In 1865, following his experimentations with spontaneous generation, Pasteur invented a method, known as *pasteurization,* for eliminating the microbes that contaminate wine, and thus avoiding serious losses to one of France's most important industries at that time (pasteurization can also be applied to other liquids, like milk, allowing them to stay fresh longer). Pasteur's prestige was so great in 1865 that the French Parliament entrusted him with the mission of saving another important sector of the French economy from ruin: the silk industry, in danger because of a plague known as *pébrine,* which had struck the silkworm. Pasteur set to work, and in three years' time, discovered and isolated the parasite that caused the disease, offering a method for ending the epidemic. In 1868, the disease was eradicated, and the French silk industry was saved from disaster.

Pasteur's career took a turn of extraordinary consequences when, in 1871, he shifted his attention from industrial processes to the field of health. Owing to his experience with microorganisms, Pasteur offered his microbian theory of disease, which holds that many diseases are caused by the action of microorganisms. This idea was received with great skepticism, but in 1881, Pasteur succeeded in proving his ideas when he developed treatments for the prevention of two grave illnesses in domestic animals: chicken cholera and carbuncle (also known as malignant anthrax). The latter disease mainly affects sheep and cattle, but it can also be transmitted to humans. It is one of the oldest known infectious diseases in history, a description of it appearing in *Exodus*. The pathogen agent, a bacteria known as *Bacillus anthrax*, was discovered in 1863 by the French scientist Casimir-Joseph Davaine (1812–1882)[1], being the first microbian disease in which the pathogen was identified. Even though in the absence of medical care the carbuncle can become life threatening to humans, treatment with antibiotics is very effective against the disease.

Again, the economic benefits of Pasteur's discoveries were enormous. At this stage, he was already recognized as one of the greatest scientists of his era and considered a national hero of France. In 1882, at the age of sixty, Pasteur began investigating one of the most terrible diseases of his time (although very low in incidence): rabies. Following his normal method, his first objective was to identify and isolate the causative agent of the disease. Once this was accomplished, Pasteur garnered samples with strains whose virulence had been attenuated and could be used to inoculate the animal. But he still needed definitive proof in humans.

The opportunity presented itself on July 6, 1885, a date on which one of the most dramatic scenes in the history of medicine took place: On that day, a woman turned to Pasteur with her nine-year-old son who had just been bitten by a rabid dog. In her desperation, the mother had come to the prestigious doctor to ask for his help, and he consented to treat the boy. The child's name was Joseph Meister, and he would be the first person to receive

1. Davaine's discovery was not accepted until 1876, when the great German microbiologist Robert Koch (1843-1910) isolated the bacteria and documented its life cycle.

Pasteur's treatment for rabies. For ten days, Pasteur inoculated Joseph with samples of the causative agent of the rabies, each time using a mixture of greater virulence. At the end of the treatment, the boy showed no signs of the disease; he was saved, and rabies was conquered.

As we have already mentioned, Pasteur's work was widely recognized during his life. In addition to his numerous monetary prizes, Pasteur received some of the most prestigious medals of his time. Among them, the Legion of Honor (in which he was awarded the rank of knight in 1853 and commander in 1868, and the Great Cross in 1881) and the Gran Cordón of the Order of Isabel the Catholic (in 1882).

But the greatest and most emotional display of gratitude for his work took place posthumously. The French people bid him farewell with a state funeral in Notre-Dame, where he was interred in one of its chapels until, in 1896, his remains were moved to a crypt in the Pasteur Institute. There, since 1910, his remains have remained in repose next to those of his wife, Marie Laurent. According to the great French microbiologist René Dubos, during the Nazi invasion, in 1940, the occupying German forces requested that the crypt bearing the remains of Pasteur and his wife be opened. The custodian in charge of the keys took his own life rather than step aside for the occupying forces. That porter was Joseph Meister, the nine-year-old boy whom Pasteur had saved from rabies.

So that what for another scientist would have meant the greatest achievement of his professional life—the solution to the problem of spontaneous generation—was simply another milestone in his career for Pasteur. Let's remember that the defenders of spontaneous generation, like Needham, believed that air was capable of generating on its own the materialization of microbes in previously sterilized culture mediums. Therefore, they ruled out experiments, like those of Spallanzani, in which the air was subjected to high temperatures, since such temperatures altered the air and caused it to lose its *life force*.

Pasteur began his investigations by demonstrating that air is not free of microorganisms. To do that, he pumped air through a glass tube, into which he had inserted a sterilized cotton ball, a kind of filter. Next, he dissolved the cotton with ether and found the remains of numerous microorganisms in the solution. These microorganisms could only have come from the air that had passed through the tube and were then trapped in the fibers of the cotton.

Once it was proved that the air contained microbes, Pasteur carried out his second, and definitive, experiment. He took a swan neck flask (this is a vessel with a very long, thin neck) and filled it with a culture medium, next bending the neck of the flask down and then up (having been submitted to heat, the glass could be easily bent). Next, he boiled the medium until vapor began coming out through the neck of the flask. The days passed, and no microorganism appeared in the medium. The experiment was conclusive; Pasteur had proved the falsity of spontaneous generation. In the words of Darwin's chief proponent, the Englishman Thomas Henry Huxley (1825–1895), written in 1863: "For my part, I believe that, with the results from Pasteur's experiments before us, we cannot avoid coming to the very same conclusions that he reached; and that the doctrine of spontaneous generation has received the final *coup de grâce*."

Apparently, Pasteur's experiment does not differ from the previous experiments conducted by Joblot or Spallanzani: boiling a culture medium inside a vessel. The fundamental difference is that Pasteur's flask *remained open the whole time*. By doing it this way,

Pasteur allowed the air in the room, which had not been subjected to any treatment at all, to enter the vessel after the medium had been brought to a boil. With that, he eliminated any doubt concerning the experiment's validity. But how is it possible to allow air to enter the flask without the flask itself becoming contaminated with microbes? The answer lies in the elbow Pasteur created when he bent the neck of the flask. Air freely passed through the neck of flask, but the particles of dust suspended in the air, where microorganisms are found, were then trapped in the elbow of the flask. The solution Pasteur provided was as simple as it was ingenious. As confirmation of his results, when the neck of the flask was broken and the elbow that retained the airborne dust was removed, the culture medium was filled with microorganisms.

Even though Pouchet tried to refute Pasteur's results, the battle was already irrevocably lost: in 1864, the Paris Academy of Sciences adopted Pasteur's results (Pouchet eschewed repeating his experiments and defended his results before the commission of

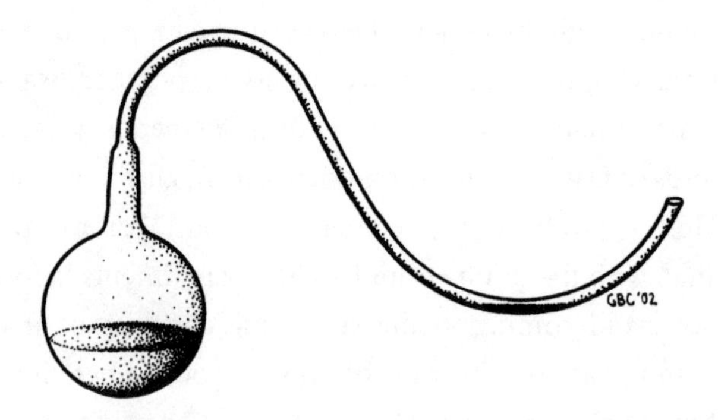

Figure 9: The swan neck flask used by Pasteur in his experiment to refute spontaneous generation. The glass neck has been bent in order to form an elbow that will retain the particles of dust in the air.

experts designated by the Academy). Sometime later, in 1869, the English scientist John Tyndall (1820–1893), the discoverer of the effect that bears his name and explains why the sky is blue, was able to observe firsthand that the putrefaction of organic matter (caused by the proliferation of microorganisms in it) did not occur in the presence of air free of suspended particles of dust, but rather when the opposite was true, thus confirming Pasteur's results.

There was still a bit of a debate regarding spontaneous generation in the 1870s. Pasteur once again entered the fray, mainly against Henry Charlton Bastian (1837–1915), who maintained that spontaneous generation was possible under very precise conditions. Again, in 1877, the Paris Academy of Sciences named a commission of experts, to which Pasteur and Bastian agreed to submit themselves. And again, as had happened ten years earlier with Pouchet, Pasteur discredited the arguments and tests of his opponent, who also on this occasion refused to directly confront him.

Nevertheless, and despite Pasteur's leading role in discrediting the theory of spontaneous generation, it is only right to recognize that its definitive abandonment was also decisively influenced by the advances in other branches of experimental biology, such as cytology. In fact, Pasteur could not explain occurrences in which, after sterilizing a culture medium through prolonged boiling, certain microorganisms inexplicably reappeared in hermetically sealed jars. Tyndall was the researcher who found the solution to this problem, on discovering that certain bacteria can develop into cysts, shutting themselves off in capsules (*spores*) that resist prolonged boiling.

When a culture medium, containing the bacteria in question, is boiled (for example, a culture medium prepared using hay as a source of organic matter, as in Tyndall's experiment), some of these same bacteria encapsulate themselves, forming spores that are resistant to boiling. When the temperature of the medium is lowered, those spores open and microorganisms reappear, apparently right where they had been eliminated. Tyndall discovered that if the culture mediums were subjected to brief periods of intense heat, separated by intervals wherein the temperature was allowed to drop to ideal levels for microbian growth, sterilization was complete, and microorganisms did not reappear in the culture medium. During periods in which the medium is heated to the boiling point, almost all of the bacteria disappear, except a few that form spores. When the medium is allowed to cool, these spores open, and the bacteria resume their normal work. If at that moment, they are once more submitted to rapid heating, only one small fraction of those few bacteria will again encapsulate themselves. By repeating the procedure several times, the total elimination of the bacteria is achieved. This technique is called the *Tyndall effect*, in honor of its inventor.

The end of the theory of spontaneous generation meant a dire crisis for scientific research on life. In barely three years (1859–1862), Darwin's and Pasteur's work had provoked a Copernican-like reversal regarding the problem of the origin of man. Darwin had shown man's relationship to the origin of other living things, and Pasteur had refuted the idea that this origin could be produced through spontaneous generation. What alternative was left for explaining the origin of life? For many, the answer lay in a divine act of creation. For those who did not want to renounce the idea that life was a natural phenomenon, that it can be explained as the result of natural causes, all that remained was the vexation over not being able to offer any explanation.

In 1864, Pasteur wrote: "Oh! In truth, at that moment, I deprived it [the substance used in the culture mediums] of *the only thing not given to man to produce*, I deprived it of the germs that float in the air, I deprived it of life, since life is the germ and the germ is life" (the cursive is ours). Why did Pasteur refer to life as "the only thing not given to man to produce?" If man cannot repeat the conditions under which life first surfaced, isn't it probably because those conditions are of a supernatural character? If Pasteur was pointing toward the act of creation, Darwin was aligning himself with those who preferred not to ask themselves a question that, at that moment, remained beyond the grasp of science. Even though Darwin, in the final paragraph of *The Origin of the Species*, had no sheepishness about attributing the origin of life to the Creator[2], he later stated, following the first edition, his frank opinion of the problem in a letter in 1863 to his friend and compatriot, the botanist Sir Joseph Dalton Hooker (1817–1911): "It is foolishness to talk about the origin of life; one could just as well talk about the origin of Matter."

But the position taken by the two nineteenth-century giants was going to change with time. In 1878, barely a year after his definitive triumph over Bastian, Pasteur wrote, concerning spontaneous generation: "I do not believe it impossible." Five years later, in 1883, he acknowledged his interest in creating artificial life. For his part, Darwin, in 1871, declared in another letter to Hooker, that life could manifest itself in a natural way "in a small, warm pond, in the presence of some type of ammonia salts and phosphoric acid, with light, heat, electricity." Both began considering the circumstances under which life could emerge from inert matter. With the crisis over, science again looked the problem of the origin of life in the face.

The Russian and The American

As we have already explained in the preceding pages, the molecules that constitute the formation of living organisms—biomolecules—

2. Although in chapter VIII, devoted to instinct, Darwin wrote, "I must state at the outset that I am no more concerned with the origin of mental faculties than I am with the origin of life itself."

are very different from the ones that comprise inorganic matter. We then said that these molecules are only found in the organisms themselves, which posed the problem of the origin of the biomolecules of the first living organism.

At the beginning of the nineteenth century, the scientific community widely held this opinion that biomolecules could only be generated by organisms themselves, and they were not spontaneously produced in nature. However, this idea was refuted in 1828, when the German chemist Friedrich Wöhler (1800–1882) was able to synthesize from inorganic salts (lead and ammonium cyanates) an organic molecule typical of living things: urea. In 1863, Thomas Henry Huxley commented on the importance of Wöhler's discovery: "It was not so long ago—and one must not forget that organic chemistry is a young science, just a few generations old, and that it would be wrong to expect much from it—it was not so long ago that it was said that it was wholly impossible to synthesize any organic compound; that is to say, any of the nonmineral compounds that are found in living things. So it was for a long time; but a good number of years have now passed since a distinguished foreign chemist was able to produce urea, a substance quite complex in nature, which forms part of the waste produced by animal groups."

Urea is a complex molecule widely distributed among different organisms. In humans, and many mammals, urea is synthesized in the liver, as a waste product of the metabolism of amino acids. When those acids are degraded, they release their amino groups, which on being set free in the blood or tissues would produce ammonia, a highly toxic substance. To avoid this, the amino group is incorporated into a more complex and much less toxic molecule—urea—that, readily soluble in water, excretes itself in the urine.

Despite Wöhler's indisputable success, which marked the birth of organic chemistry, scientific investigation into the origin of life still ran into difficulties with two serious problems in the nineteenth century. On the one hand, knowledge of organic chemistry, in general, and biochemistry, in particular, was minimal (just as Huxley pointed out in the words that were quoted), and thus limited putting into place detailed hypotheses on the chemical processes that could have originated life. On the other hand, there was another question that was also hard to answer: if organic matter could be generated from inorganic matter, without the participation of living things, why were such processes not being repeated in the present?

For some scientists, the answer to this question was that life did not originate on Earth, but came from some other place in the universe where physical and chemical conditions were more favorable to their emergence. One of the principal advocates of this hypothesis was the Swedish Nobel Prize winner for chemistry, Svante Arrhenius (1859–1927). The idea of extraterrestrial life found little support (even though among its defenders were scientists of the stature of Harry Francis Compton Crick, Nobel Prize for medicine, the discoverer, along with James Dewey Watson, of DNA's tridimensional structure), basically because the problem lies outside the reach of experimental science, and because it is possible to formulate alternative theories that clearly place the origin of life on our planet.

In order to understand what circumstances must be present for life to originate on Earth, we have to come back to the chemical composition of our world. The current terrestrial atmosphere is dominated by the presence of nitrogen (77 percent) and oxygen (21 percent). The vast quantity of free oxygen in the atmosphere accords it the nature of an *oxidizing* agent, something incompatible with the chemical reactions that must occur for organic matter to combine and produce inorganic matter, which is greatly *reduced*. Later on, we will explain in more detail the concepts of oxidation and reduction, but for now, all we need to know is that an atom is oxidized if it is combined with the oxygen atom and reduced if it is combined with the hydrogen atom.

The oxidizing nature of Earth's current atmosphere impedes the spontaneous emergence of biomolecules from inorganic compounds; however, the atmosphere has not always had the same composition. In the first place, the huge amount of free oxygen in our present-day atmosphere is remarkable. Oxygen is highly reactive and tends to bond (oxidize) with the other chemical elements, and for that reason, should not be found free, but rather combined with other elements in the form of oxides. The only explanation for the high concentration of oxygen in the current atmosphere is the action of photosynthesizing organisms, which break apart the molecules of water (H_2O), freeing the oxygen. In other words, the presence of oxygen in the atmosphere comes after the appearance of life. It then seems clear that the atmosphere, in which the chemical reactions that took place and produced life, lacked oxygen and was not as oxidizing as it is now.

On the other hand, it is quite reasonable to suppose that the primordial atmosphere was much richer in hydrogen than it is today. Hydrogen is the most abundant element in the universe, and it is natural to think that it was also a major component in the primitive atmosphere. On small planets, like Earth, the force of gravity is not able to retain small atoms—hydrogen, being the smallest of all—which gradually escape into their atmospheres over the course of time. Contrariwise, large planets, like Jupiter, have a gravitational field sufficiently strong enough to retain light atoms, and therefore, their atmospheres maintain a chemical composition more like the initial one. This is why Jupiter's atmosphere is richer in reduced molecules, like *methane* (CH_4) and ammonia (NH_3). Taking Jupiter's atmosphere as a model of what could have been Earth's primordial atmosphere, it has been theorized that our planet lacked oxygen and was rich in reduced molecules, like methane and ammonia.

Among those who believed this was a Russian biochemist, Alexandr Ivanovich Oparin (1894–1980), who, in 1922, presented a work to the Moscow Botanical Society, suggesting that the core of an atmosphere of those characteristics (devoid of oxygen, but rich in water, methane, and ammonia), submitted to intense ultraviolet radiation and jolted by electrical charges, would have produced a series of chemical reactions that in turn would have resulted in the appearance of all the different organic molecules. According to Oparin's theory, these molecules accumulated in the ancient oceans and, in turn, probably continued reacting and combining to produce more complex molecules each time. Finally, these complex molecules separated from the medium, by means of chemical-physical processes, in vesicles (*small bladderlike cells*) from which the first living organisms emerged.

Oparin's hypothesis benefited from the vast body of knowledge of organic chemistry that was assembled in the first decades of our century. By that time, numerous organic compounds had been artificially synthesized, and the chemical bases of those procedures were well known. Oparin also based his work on the development of investigations into the origin of Earth and the possible composition of its primal atmosphere. But in addition to this knowledge, Oparin's theory was greatly influenced by the ideas of the German philosopher, and one of the fathers of socialism, Friedrich Engels (1820–1895), as well as by those of Darwin himself.

From the former, Oparin took the idea that matter evolves over time, acquiring ever more complex forms that result in living matter. This idea is crucial, since it allows us to believe that the existing abyss between inorganic matter and living organisms can be bridged in a series of stages of growing complexity, and not in a single leap. But the chief problem in understanding the origin of life, as springing from a series of chemical reactions of inanimate matter, is based on the fact that the aforementioned chemical reactions are produced in a deviating and random fashion. How could those reactions link in just the precise order, all by themselves, to produce something so delicately arranged and adapted as living matter? Or to put it in simpler terms, how can order emerge from chaos? Oparin found the answer in Darwin's stroke of genius: natural selection.

In a celebrated quote, the great Albert Einstein (1879–1955) clearly stated his opinion on the role of chance in the natural world: "God does not play dice." In Darwin's time, many of his critics, especially from the field of paleontology, made the same criticism of his theory of evolution (his and Wallace's). Those scientists could not imagine that the extraordinary design living things displayed in all their structures might belong to a process in which, at its very heart, pure chance was at work. However, those criticisms were without foundation. Darwin's real discovery, natural selection, is a mechanism that *directs* the action of chance, choosing what is best (the best adapted) among those randomly produced variants. Darwin's dice are not the ones referred to by Einstein; Darwin's dice are charged, souped-up by natural selection; despite the many possibilities, only some of them will manage to prevail.

Thus, according to Oparin, the chemical evolution of matter produced small bladderlike cells—vesicles—filled with haphazardly joined organic molecules, which were selected based on their ability to survive. In this way, of all the vesicles that were formed, the only ones to last were those that contained particular combinations of biomolecules whose chemical reactions conferred on them greater stability. From among all those that continued to predominate, the ones that triumphed were those capable of maintaining, over a long period of time, the suitable combination of organic molecules, exchanging matter and energy with the surrounding medium. Finally, from among the chosen cells the ones were selected that had acquired the ability to increase in number once they were able to diffuse away from each other without any loss of their properties.

It is relatively frequent in the history of science for two scientists to make the same discovery, or propose the same theory, independent of each other and almost simultaneously. After first stating his hypothesis in 1922, Oparin published, in 1924, a book of modest scope in which he gave a detailed description of his theory. Later, in 1936, his best known and most extensive work appeared: *The Origin of Life on Earth*. Independently in 1928, when Oparin's work was still unknown in the West, one of the most versatile scientists of the twentieth century, John Burdon Sanderson Haldane (1892–1964), published a theory on the origin of life very similar to Oparin's. Because of it, both authors are often linked together, and their work spoken of as the Oparin-Haldane theory.

For a quarter of a century, Oparin's ideas (and Haldane's) were the subject of controversy, attracting support and raising rebuttals, but the argument was strictly maintained on the theoretical level. Then, at the beginning of the 1950s, a young American thought about carrying out an experiment to test the Russian scientist's fundamental theory. The name of this young man was (and remains today) Stanley Miller; he was an unpretentious doctoral student, merely twenty-three years old, at the University of Chicago, when he conceived and brought to fruition a simple experiment, whose results comprise one of the greatest scientific discoveries of the twentieth century.

Before explaining Miller's experiment and its importance, we should comment on one aspect concerning the circumstances under which he carried it out. At the time he performed his experiment, Miller was a student in the chemistry laboratory of Harold Clayton Urey (1893–1981), awarded the Nobel Prize in chemistry in 1934 for the discovery of deuterium (you remember deuterium, the hydrogen isotope with a single neutron in its nucleus) and one of the fathers of the atomic bomb. Urey was also a specialist on the subject of the chemical composition of Earth's primal atmosphere, and his opinions were similar to those of Oparin, in that the primal atmosphere lacked oxygen and was rich in methane, ammonia, and water. Urey's students were quite familiar with his ideas, since he defended them in the seminars he taught. It was none other than Urey's lectures that induced Miller to mount his experiment.

Most authors who describe the experiment in question are very careful to point out that Miller performed it under Urey's watchful direction. However, another version was told by the French scientist Joël de Rosnay, who stated that Miller's idea, of recreating the conditions Oparin had proposed regarding the primordial atmosphere and testing to see if biomolecules appeared, was especially daring since, under the aforementioned conditions, an enormous number of compounds, whose analysis was outside the reach of a mere doctoral student, could be formed. For that reason, Miller had probably not performed his experiment under his director's attentive eye, but rather in secret.

Even though the story told by de Rosnay is more amusing than the *official* version of Miller's experiment and fits quite well with a certain mythology regarding student-professor relationships, it is assuredly false. Fortunately, in 1974, Miller described the circumstances that surrounded his work and relationship with Urey. In fact, when Miller proposed the idea of an experiment to test Oparin's hypothesis, Urey expressed skepticism. In Urey's opinion, the experiment Miller suggested could produce a great number of organic compounds, whose analysis would require tremendous time and effort. Moreover, and this was the main obstacle Urey saw, the results might not prove conclusive one way or the other. The professor proved reluctant to involve his student in a doctoral project that would be difficult to execute and have very doubtful results. Rather than do that, he suggested another research topic that assured Miller of successfully completing his doctoral dissertation.

But Miller, with the typical stubbornness of youth, insisted that his original idea (the experimental testing of Oparin's thesis), and not some different one, was the topic he wanted to tackle in his dissertation. Faced with Miller's insistence, Urey proposed allowing him six months to a year to test his idea. If in that time, he did not secure definitive results, he would have to abandon his plan and accept the doctoral topic being proposed. Miller agreed and *between the two of them*, they designed the experiment and apparatus for carrying it out. Miller set to work, and after some preliminary experiments, in barely three and a half months, he obtained manifest and spectacular results.

Then an episode occurred that illustrates Urey's authentic moral standing. Convinced that his results carried great importance, Miller asked Urey's permission to write an article and send it to *Science* magazine. Urey encouraged him to do it. Once it was complete and before sending it, Miller showed it to Urey for his approval. What happened next was so exceptional that it deserves letting

Miller tell it in his own words: "At the time, I wrote a brief article, listing Urey as co-author, and took it to him for his approval. The first thing he told me was to withdraw his name from the article because I had performed most of the work myself, and if his name appeared in the article, I would receive little or no credit. That was extraordinarily generous on his part, because in chemistry the director of research is almost always included in articles that result from student dissertations."

Of what did the experiment consist, and what were the results? In the laboratory, Miller recreated the physical and chemical conditions that Oparin and Urey believed were Earth's primitive atmosphere. Miller built a closed circuit consisting of a narrow flask placed beneath a large, bell-shaped vessel and connected to it by glass tubing; two electrodes were placed inside the flask, and another tube was affixed to the bell-shaped vessel that came out at the top and ran back down to the first flask. Next, the tubing was bent to form an elbow.

After testing the vacuum inside the circuit, the experiment began when Miller placed a mixture of water, hydrogen, ammonia, and methane inside the flask. Next he boiled the water inside the flask, generating a current of warm air and vapor, and causing the mixture of water, hydrogen, ammonia, and methane (which represented the hypothetical primitive atmosphere) to circulate through the closed circuit for days. With this mixture passing through the chamber containing the two electrodes, it was periodically submitted to electrical charges of 60,000 volts, which simulated the lightning of Earth's original atmosphere.

After a week, an orange liquid had accumulated in the elbow of the glass tubing. Miller proceeded to make a meticulous analysis of the orange substance and found in it numerous organic molecules, among them urea and, above all, some of the amino acids present

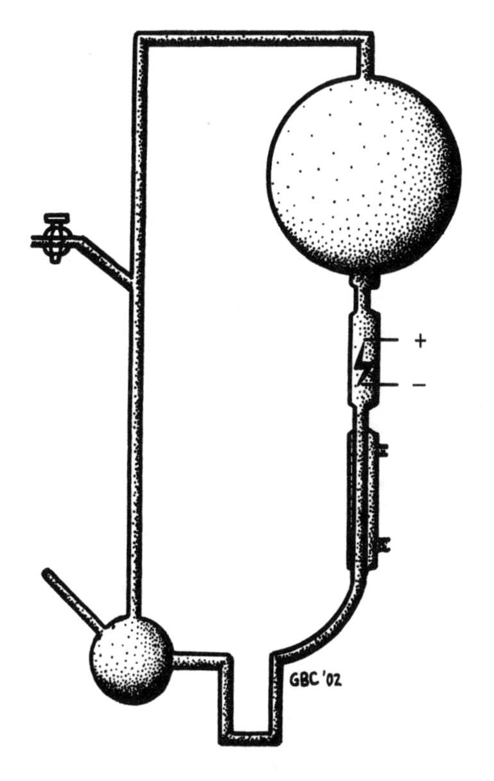

Figure 10: *A diagram of the apparatus devised by Miller for his experiment with the synthesis of organic matter under the theoretical conditions of Earth's primitive atmosphere. The + and - symbols indicate the position of the electrodes that administered the 60,000 volts of electrical charge.*

in living organisms. Standing before what he expected to see, at least in theory, based on the experiment's rudimentary conditions, Miller, instead of finding formations of an extraordinary number of very diverse molecules, found only a limited number of them, among which were some of the essential biomolecules in living things. Miller had spectacularly confirmed Oparin's ideas: under the theoretical conditions of Earth's primitive atmosphere, the building blocks of life had materialized from inorganic matter.

Miller published his results in *Science* magazine in 1953, the same year that Watson and Crick published the DNA's tridimensional structure in the journal *Nature*. Without a doubt, 1953 was a wonderful harvest for biology.

Miller's experiment was a boon for research into the origin of life and opened the way for its experimental study. Ever since then, a great many experiments have been carried out, varying the initial conditions, either the original molecules or energy source. As a result, all kinds of biomolecules of living organisms have been experimentally obtained. Some of the most resounding successes in this research belong to Juan Oró.

In 1960, Oró mixed ammonia and hydrocyanic gas with water and heated them to 194°F, conditions compatible with those of primitive Earth, obtaining adenine. You will remember that adenine is one of the components of nucleic acids (DNA and RNA) and ATP (adenosine triphosphate), fundamental molecules in living organisms. The finding was so unexpected that, initially, Oró thought he had made a mistake. Later, in 1963, he succeeded in synthesizing, also under conditions similar to those of Earth's primordial atmosphere, ribose glucides and deoxyribose, the other constituent biomolecules of DNA, RNA, and ATP.

However, in recent years, these successes were overshadowed when doubts mounted regarding the original composition of the terrestrial atmosphere being so favorable to the manifestation of the different bio-molecules. There are facts suggesting that the reducing primitive atmosphere was soon replaced, before the conditions conducive to the emergence of life materialized, by another of neutral character. Under these conditions, the acquisition of biomolecules is not as successful. While the debate goes on regarding the nature and evolution of Earth's primitive atmosphere, other possible scenarios have been suggested with respect to the emergence of biomolecules, even in a nonreducing atmosphere.

One of them points to volcanic types of environments, like fumaroles on ocean floors, where high temperatures, as a source of energy, and reduced molecules are produced at the same time. The second possibility, presented by Oró in the 1960s, looks to the sky and even outer space once more.

The results of Miller's experiment received widespread support with the discovery of biomolecules of terrestrial origin. For many years, scientists knew of the existence of organic molecules (amino acids, nitrogenous bases, and lipids) in a special type of meteorites known as *carbonaceous condrodites*. For some time, the origin of these biomolecules was unclear, since contamination from Earth's atmosphere could not be ruled out. However, one fortunate event came along to dispel the doubts. In 1969, a meteorite, of the type cited above, fell in Australia, near the Murchinson gold fields, and was analyzed shortly thereafter, before any contamination could take place. The results of the analysis are very revealing: six of the twenty amino acids used by living organisms were found. These amino acids had also appeared in Miller's experiment in similar quantities. Thus, Miller's experiment has a universal validity: wherever proper conditions are created, the molecules of life will appear. From this point of view, the question of whether these life molecules were formed on Earth, or came from outside, is of lesser transcendence. What is really important is that inorganic matter can *spontaneously* generate organic matter.

In recent years, we have been discovering that the presence of biomolecules in outer space is not an unusual occurrence, but rather a relatively frequent one. This is confirmed by the analyses conducted on meteorites preserved from terrestrial contamination during thousands of centuries under the Antarctic Circle. We have also detected, thanks to radio astronomy, the presence of biomolecules in clouds of galactic dust and interstellar space. In 1988, the

Giotto comet chaser of the European Space Agency crossed the tail of *Halley's* comet, revealing an unheard of amount of organic matter.

The first step has already been taken. The formation of biomolecules from inorganic matter is an indisputable fact nowadays. Yet to be explained are the following stages on the road to life: how could organic molecules spontaneously organize themselves to produce the first living organism? At this point in the research, there are numerous approaches and different opposing models have been formulated. But we still lack experimental results that will allow us to map out the field of possibilities in a meaningful way. Perhaps we are close to realizing an experiment (maybe it is being carried out at this very moment) that, like Miller's, sheds light on our understanding of the origin of life.

Until that moment arrives, let's again concede the word to Miller, who, in answering the question as to whether the origin of life is an enigma within reach of the knowledge of man, said: "I believe that we have not yet found the right gimmick [. . .] When we find the answer, it will be so wretchedly simple, that we will all be asking ourselves, how come I didn't think of that sooner?"

Are We Alone or Do We Have Company?

The majority of public opinion is firmly convinced of the existence of extraterrestrial life. So strong is the certainty of its existence that it has practically been transformed into a dogma of faith, fitting for our times. When someone claims to have witnessed the appearance of the Virgin, skepticism, if not ridicule, is the most frequent reaction among the media. It doesn't matter how many witnesses there were at the sighting, or that there were highly educated people among them; incredulity is the norm. This situation stands in stark contrast to the fuel given to the sighting of alien spaceships, and even personal and direct contact with their crew members. Movies and literature of our time are so filled with extraterrestrial beings that it almost seems strange not to see them on the street.

If anyone publicly decides to place in doubt the existence of life outside our planet, he runs the risk of being treated as an ignoramus and hearing such authoritative comments as: "Of course, life exists, but different from the one we know!" or "How is it possible we're the only living beings in the universe!"

The first argument is always amusing because if extraterrestrial life is really so different, how would we recognize it, even were we to trip over it? In fact, the only possibility we have of recognizing the existence of life outside our world lies in its *looking substantially like* life as we know it. A historical example exists of our ability to recognize living beings such as they are even if they are very different from the ones that are familiar to us. When Leeuwenhoek discovered microorganisms, in 1683, he had no qualms about classifying them as living organisms, despite the fact that they vastly differed from the known organisms up to that time. Beyond the great difference that existed between microbes and animals, Leeuwenhoek perceived a characteristic that he believed common to animals: autonomous movement.

Today there is a branch of biology that seeks to determine what those basic characteristics of life are, with the goal of finding and identifying it outside our planet. It is *exobiology*. Jokingly, it is routinely said that exobiology is the only science without any purpose to its study, since there is no known extraterrestrial organism. However, Cuvier, the father of paleontology, could be given the honor of having been the first and only (up to that time) exobiologist in history, since he had the opportunity to analyze a being not of this world.

Cuvier was the creator of one of the most important conceptual instruments for the study of fossils: the *principle of organic correlation*. This principle holds that the different parts of living organisms are so intricately arranged, functionally and structurally, to one another, that it is possible to deduce the modus of life and the complete structure of an organism from some of its parts. Taking a position based on this principle, Cuvier amazed the society of his era by reconstructing lifeless organisms from a few fossils. Today, we paleontologists, consciously or not, continue using Cuvier's principle of organic correlation in our reconstructions. Cuvier's academic career was meteoric, and around 1800, at only thirty-one years of age, he was a professor at the College of France. But his scientific and academic brilliance was not accompanied by personal good fortune. It is said that his students hated him. Apparently, as Herbert Wendt tells it, in his book *In Search of Adam*, Cuvier's students one day decided to get back at their detested teacher, playing a practical joke on him. One night, they slipped into his room, and one of them, fittingly disguised as the devil, approached the sleeping Cuvier, saying something to the effect of: "Wake up! I'm the Devil, come to devour you." Hidden in the shadows of the room, the rest of the students were licking their lips in anticipation of the terrible scare that Cuvier was about to receive. But the professor, not the least bit frightened, opened his eyes and coldly replied: "Impossible, you've got horns and hooves. According to the principle of organic correlation you've got to be herbivorous." After a second of silence, a round of applause arose from the darkness of the room.

Fact or fiction, this anecdote illustrates a really interesting point for us: paleontology is also faced with the problem of studying extraterrestrial living organisms. Time, not space, separates the organisms we know of today and are the subject of paleontological study. Like exobiology, paleontology also attempts to understand the basic characteristics of life in order to be able to use them in its studies.

But let's return to the matter of extraterrestrial life. The second argument ("How can we be the only living organisms in the universe!") has more substance and can be pondered from a scientific perspective.

In 1961, the U.S. astrophysicist Frank Drake constructed an equation, known as the *Green Bank Formula,* to calculate how many extraterrestrial civilizations might reasonably exist in our galaxy at this time. But since our interest in this question goes beyond the present moment, we can modify that equation to calculate how many extraterrestrial civilizations have come into being in the entire history of the galaxy. From this perspective, the *Green Bank Equation* (modified) reads like this:

$$N = R x f_p x n_e x f_l x f_i x f_c$$

Don't be intimidated; the equation is quite simple. N is the total number of civilizations that have existed in the galaxy since its formation. R is the total number of stars formed in the galaxy. The f_p factor is the fraction of the said number of stars in the planetary systems tracing orbits around them. The average number of planets that can support life in each planetary system is represented by the n_e coefficient. The probability of life emerging on a planet with favorable conditions is f_l. The coefficient f_i represents the probability that intelligence will emerge, and f_c is the probability that intelligent life will develop a technological civilization. Since what interests us here is the existence or nonexistence of extraterrestrial life, we will forego the last two coefficients f_i and f_c.

According to astronomers, the approximate number of stars in our galaxy (R) is close to 2 x 10^{11}. Apparently, it is reasonable to suppose that half those stars contain planetary systems ($f_p = 0.5$), and that on average, each system includes two planets with adequate conditions for the emergence of life ($n_e = 2$). This leaves us with the astonishing number of 2 x 10^{11} planets that could support life. To give you an idea of what that number represents, think of it as a little more than the distance from Earth to the sun, expressed in meters, or the number of seconds there are in 6,342 years.

Perhaps these calculations seem a little too optimistic to you. It's possible, but there's a reason for it. In the first place, the fact that the stars have planets in orbits around them is an inevitable consequence of the process of their formation. The star with which we're most familiar, the sun, has its planetary system, and we know of more examples in the galaxy; in which case, it is not ludicrous to suppose that half of the stars can have planets orbiting them.

Nor does it seem an exaggeration to say that each planetary system has, on average, two planets suitable for life. There isn't any reason to assume that the planets of the hypothetical planetary systems have to be concentrated in the vicinity of their corresponding star, where the temperature would be quite high, or situated very far from it, which would imply temperatures that are too low. The reasonable thing to believe is that the process of star formation, and of their accompanying planets, produced planets of different masses, located at different distances from the mother star, some near, others at intermediate distances, and others farther away. It is also reasonable to assume that a good number of these planets have sufficient mass to generate a gravitational field capable of retaining an atmosphere. Under such circumstances, two planets capable of generating and accommodating life do not strike us as an absurd number.

Still, if that is true, why does our solar system, the only one that we intimately know, only contain one planet, Earth, with life? Isn't it probable that, in the end, the calculations are too optimistic? What is true is that there are at least two planets in our solar system capable of supporting life. These planets are Earth . . . and Mars.

The red planet is situated somewhat farther beyond the sun than the blue planet, but the difference is not relevant as far as the conditions for supporting life. A more important difference lies in the size of our cosmic neighbor. Mars is appreciably smaller than Earth; as a consequence, its gravitational field is weaker and atmosphere more tenuous. This, indeed, is a serious problem, but not unsolvable, for the development of living beings. A thin atmosphere allows a large quantity of ultraviolet radiation to reach the surface. However, in its liquid state, water can serve as an efficient filter for ultraviolet rays, and life could take refuge below its protective layer.

Unfortunately, on Mars, there is no water in a liquid state. The planet is too cold, and its water lies frozen. But this is the current condition; there are very firm indications that Mars enjoyed more benign temperatures early on, and water freely circulated on its surface. In which case, life could have appeared on Mars and may yet remain under the layer of ice of its polar icecaps. In fact, bacteria and unicellular algae have recently been found on our planet, living under similar conditions under the Antarctic ice. Maybe those hypothetical Martians (surely microbes) are extinct, but there will always be the possibility of finding traces of their existence. This probability opens the field to a new science: *exopaleontology*.

That Mars is considered a prime candidate for supporting (or having supported) life is proven by the fact that two spaceships have been sent there, *Vikings* 1 and 2, suitably equipped with the most refined devices for detecting the presence of living things. Both reached the Martian surface in 1976, in a matter of six weeks, at separate points a little over 4,000 miles apart, and carried out a series of experiments designed to detect the presence of life on Mars. Initially, the results obtained in those experiments seemed to point to the existence of life on Mars, but more detailed analyses led to the conclusion that an error in interpretation had been made . . . although a shadow of doubt will always remain.

In 1996, hope of finding life of extraterrestrial origin was rekindled. In August of that year, a finding was published that in the interior of a modest meteorite (weighing somewhat less than two kilograms), recovered in 1984 from the Antarctic ice and bearing the abbreviation ALH84001, they had found carbonate nodules, whose chemical composition included molecules and organic structures similar to those generated by terrestrial bacteria. The ALH84001 meteorite reached Earth thirteen thousand years ago, torn loose from the Martian surface 16 million years earlier by the impact of a large asteroid that had struck the planet. Unfortunately, although the Martian origin of ALH84001 is indisputable, the biological nature of the professed biomolecules is quite debatable. Following the initial clamor, a substantial segment of the scientific community does not consider the evidence found in ALH84001 to be an unquestionable proof of alien life.

In any case, whether there was life on it, Mars seems to be *the other* habitable planet that statistically matches up with our solar system.

Up to this point, we have discussed that part of the *Green Bank* equation that concerns the inorganic world. But now it is a question of establishing on how many of the potentially fertile planets life might have originated. In order to do that, we have to ask ourselves what is the probability of life appearing when all the circumstances are in its favor (the coefficient f_l of the equation). If, as some authors conjecture, life is a regular occurrence of matter, then the supposed probability will take on a value of close to one (it will occur quite frequently), and the universe will bubble with life. But if, as others think, life is a singular event, then the probability that it will appear is indistinguishable from zero (it will almost never occur), and we will have to get used to the idea that we are alone.

The champion of the most pessimistic view was Jacques Monod who, in his book *Chance and Necessity*, expressed the conviction that life is a phenomenon of such low probability that it makes no sense to hope that it has happened anywhere else or at any other moment: "The ancient covenant is in pieces; man knows at last that he is alone in the universe's unfeeling immensity, out of which he emerged only by chance."[2] According to Monod, the phenomenon of life is a singular event of natural origin; that is to say, it is compatible with the laws that govern the universe, even though it is not an inevitable, not even a predictable, outcome of those laws. Life is only one out of the many possible events with an a priori probability of being produced null and void.

But alternative views to Monod's do exist, and they come from the field of molecular biology and the branch of physics that studies the thermodynamics of systems far removed from equilibrium. Beginning with the first of them, theoretical models have an experimental basis and downplay the role of chance upon which the road to life have been speculated. In short, these models postulate that from a group of simple biomolecules, a phenomenon of evolution can be established, governed by natural selection, which leads to the appearance of complex molecular systems with some of the basic characteristics of living things: chemical-physical stability, efficient self-replication, and an ability to evolve over the course of time.

2. Translated from the French by Austryn Wainhouse (Vintage Press, 1971).

On the other hand, it has been discovered that in thermodynamic systems quite far removed from equilibrium, matter is spontaneously organized in very tidy systems, called *dissipative structures*. A system is in thermodynamic equilibrium when its energy and matter are homogeneously distributed within. This homogeneous distribution is the system's most disordered state possible, whereby the second principle of thermodynamics assures us that all closed systems will tend toward the state of equilibrium because it is the most disordered (or, if you remember, of greatest entropy).

Let's look at a simple example: if you fill the bathtub with cold water, and then introduce a hot-water bottle into it, you will have created a situation of thermodynamic disequilibrium, since the heat is not homogeneously distributed throughout the system, and one zone will be warmer than the other. But this situation is short-lived, as the cold water absorbs heat from inside and outside the bottle. The heat was homogeneously distributed, the thermodynamic equilibrium was reestablished, and the second principle records a new victory.

However, there are situations in which systems are far removed from a permanent state of equilibrium. Oceans are heated differently at the equator than in arctic regions. Inexorably, owing to Earth's spherical form, solar radiation maintains the oceans in a thermodynamic situation far removed from equilibrium. In this situation, we witness a marvel: water is spontaneously set in motion, and rivers materialize in the heart of the ocean. These are oceanic currents that carry warm water to the cold regions and frigid water to the warm regions. These oceanic currents are very organized structures with billions of molecules moving in unison on a determined course. Oceanic currents are a very simple example of a dissipative structure, generated in a system far removed from equilibrium. Evidently, the existence of these currents accelerates the process of heat exchange between cold and warm regions, and therefore, they serve to reestablish thermodynamic equilibrium.

Our planet receives an enormous amount of energy from the sun. A large part of this energy arrives in the form of light, a form of *high-grade* energy. A large part of this energy, in turn, is beamed into outer space, and the rocks and water on our planet's surface absorb a small portion, warming itself in the process. At night, the objects heated during the day return to outer space the energy that was absorbed in the form of infrared radiation, a form of *low-grade* energy. In the process, light (a very *ordered* energy) is converted into heat, the most *disordered* form of energy.

If we stop to reflect on what was previously said, we will reach the conclusion that the terrestrial surface, inundated with light, is a system very far removed from equilibrium in relation to dark outer space. Under these circumstances, it would be legitimate to expect the appearance of dissipative structures, which accelerate the process of converting light to heat. Clearly, these dissipative structures exist: they are living entities. Photosynthesized organisms absorb a great quantity of light from the sun, which otherwise would be beamed back into outer space. The light trapped by those organisms is transformed into organic matter, which, in turn, is used as fuel for acquiring the energy with which we living beings escape the tyranny of the second principle. In this process, all light captured in photosynthesis is converted to heat. In this way, the biosphere (the aggregate of living things) is kept alive by transforming light into heat, accelerating the increase of disorder in the universe.

Focusing on the problem from this perspective, which contemplates living things as dissipative structures, life emerges as a phenomenon not only compatible with the laws of physics, but almost as an inevitable event. In the words of the Belgian scientist, of Russian origin, and Nobel Prize winner for physics, Ilya Prigogine: "On the contrary, it does not seem ludicrous to consider the phenomenon of life to be as predictable as the crystal or liquid state."

As one can see, there is no agreement among scientists on the nature of life, either as an unrepeatable, chance event or as an inevitable, normal event in the universe. However, in all honesty, it is the second of these two possibilities that enjoys the most adherents today.

We're not molecular biologists, biophysicists, or specialists in thermodynamics, and therefore we won't get into a discussion of those questions. But there is one aspect of the problem, pointed out by Monod, which concerns us as students of past life. If life is the inevitable result of the chemical-physical properties of matter, shouldn't it have emerged more than once on a planet as hospitable as ours? The universality of the genetic code guarantees us that all living entities in our world proceed from the same form of ancestral life.

Of course, the conditions of our world have changed a great deal since life first appeared, and not only regarding chemical composition or physical conditions. Just as Darwin already indicated (in the same letter to Hooker in 1871 that we mentioned earlier), the very existence of living things impedes the appearance of new forms of life: "Today, matter as we know it, would be instantaneously devoured or absorbed, which would not have been the case before living creatures were formed." However, as Darwin also points out, this argument is not applicable to the initial moment, when there were still no living organisms on the face of Earth. Under those conditions, shouldn't several forms of life have originated, different from the one we presently know, at least in their genetic codes?

For the biomolecules that formed in the primitive atmosphere to have reacted with each other on the path toward life, having continuously fallen on the surface of the planet, two conditions would need to have been fulfilled: first, they would have had to dissolve in a liquid medium, and second, they would have had to exist in high concentrations within that medium. If the concentration was low, the probabilities of them being found and being able to react to each other are drastically reduced. Therefore, it has been suggested that one of the ideal mediums for the appearance of life would be a pond, where the evaporation of water would create the adequate concentration of molecules. According to this model, there would have to have been a great number of these ponds, isolated from each other, with each of them representing a different environment. Under those conditions, the process for the appearance of life could have been produced. If life is a high probability event, it had to have appeared several times in some of those ponds. Where are those different forms of life, and how is it that they have not reached us?

It can be argued that perhaps they did appear, but that a later process of competition among them only selected the one that we know today, it being the most suitable of all. This argument conflicts with what we observe in the current biosphere, where competition among the species, the disparity of mediums, and geographical isolation have produced and maintained throughout time an extraordinary biological diversity. Furthermore, we are familiar with numerous cases of *living fossils* (including very primitive types of bacteria) that are representative organisms of biological groups that flourished in the past and were then replaced by other better adapted species after a process of competition. Can anyone believe that only one type of atmosphere existed on Earth, common to the planet's entire surface and clearly favorable to the only type of life that we know today? Can anyone defend the idea that back then, geographical barriers did not exist among the numerous places on Earth, ponds or oceans, where life could be generated, and that a phenomenon of competition was produced on a planetary scale of such devastating consequences that it totally wiped out all forms of life different from the current one, without leaving a single trace of any of them in any corner of the planet? The alternative explanation is to accept the fact that we only know one form of life on our planet because it originated but once, at a specific time and place. In other words, a fortunate event.

As you can see, there are arguments to everyone's liking. Only when the day comes when a living organism is generated in a laboratory, by means of a repetitive mechanism, or we find evidence of extraterrestrial life, will we have resolved the dilemma in favor of those who believe that life is a high probability event in the universe. Meanwhile, until that day comes, we have already moved a long way toward an understanding of the process that originated the first living being. Perhaps we can't even know how much chance and necessity had to do with the origin of life, but if there is one thing of which we are certain, it is this: it is not necessary to invoke supernatural causes to explain it. Whether it was a predictable or chance event, life is compatible with the laws governing the universe.

As for our own opinion on whether life is a phenomenon of low or high probability, and having expressed it in the form of a reply to the question that provides the title for this chapter, we do not expect to find many neighbors in this quarter.

• • •

Dawn

Our planet is a world that is continually changing its hard crust. Beneath the first few kilometers of solid rock (the *lithosphere*), the heat generated by the radioactive disintegrations that take place in its core creates a thick layer of semimolten rock (the *asthenosphere*). The rigid outer layer slides above this mantle of fluid rock, causing the movement of continental masses, joining and separating them. At certain spots on the globe's surface, the hot material of the planet's interior escapes its confinement and, once in contact with water or air, becomes solidified, forming a new lithosphere. There are also places where a good part of the cold, hard lithosphere sinks (*subsides*) toward Earth's interior, where it will again partially melt.

The rocks that form the continents are lighter than those that form the oceanic floors, and therefore, the continental masses always remain *afloat*, without being dragged down toward the deepest zones of the planet's interior. But the surface of the continents is being implacably denuded by wind and, above all, water, propelled by the energy of the sun. The sediments left by this action accumulate in certain regions and become compacted, thus forming new rocks: sedimentary rocks.

The life of sedimentary rocks isn't placid either. Submitted to the colossal thrust of the continents, in their eternal shifting, they are folded and fractured, thus creating a good part of the mountain ranges of our world, upon which the indefatigable weathering of wind and water are again brought to bear. On other occasions, the sediments are dragged together with the oceanic floor where they are deposited toward the red-hot abysses of the earth's interior, where they melt and remain until they can return, in the form of lava, to the planet's surface to begin a new cycle.

It is easy to understand that, subjected to this incessant activity, the lithosphere has been recycled many times since our world was formed, and that it no longer contains any of the rocks that constituted its primitive shell. Despite that, scientists are very sure that our planet was formed more than 4.5 billion years ago. This figure is theoretically deduced from the process that is believed to have given origin to our solar system; according to those calculations, the planets must have already been formed between 4.5 and 4.7 billion years ago. However, in the twentieth century, during the '70s, the oldest known terrestrial rock was only 3.8 billion years old, which left an enormous lapse of time of more than 700 million years with respect to the estimates of theoretical models. But we have already said that we shouldn't count on finding any rock in the current terrestrial lithosphere dating back to the time of the origin of the planet. Under those conditions, can the model be tested? The answer can be found in our neighbor, the moon.

The moon is an inert planet, whose small size causes its internal heat to quickly dissipate in outer space. As a consequence, the lunar interior is uniformly rigid, devoid of fluid layers. On the other hand, since the moon lacks an atmosphere and water in a liquid state, terrestrial phenomena, such as erosion, are not produced. Because of this, rocks on the lunar surface, unlike those on the surface of Earth, have not experienced recycling. It is indeed possible to find rocks on the moon as old as the planet itself. But when did the moon originate? Did it happen before or after the formation of Earth?

Three hypotheses have been proposed, explaining the formation of our satellite. The first one supports the idea that it originated in another place in the universe and was then captured by Earth's gravitational field. This concept has been abandoned today because it is very difficult to explain, physically, how the capture and stabilization of such a large celestial body (in relation to the size of our world) could have been produced in an orbit so close to our planet.

The second possibility is that the moon independently originated, in the same place it now occupies, but through a process similar to the one that produced Earth. In this case, both planets would be very similar in age. However, the composition of the moon is hardly compatible with this hypothesis. If it had originated in the same place and through an identical process as Earth, the moon must have had a composition similar to that of the blue planet. But this isn't the case, since the moon is less dense than Earth, which suggests a different composition.

Curiously, the hypothesis that enjoys almost universal acceptance today was originally proposed by Darwin's second son, George Howard Darwin (1845–1912), in the last decade of the nineteenth century. According to this theory, the moon is an offshoot of Earth. In his first formulation, this hypothesis holds that the centrifugal force generated by the terrestrial rotation could have caused the detachment of part of Earth's mass before it had solidified. Today, however, it is more likely thought that a collision took place between primitive Earth and a gigantic celestial body, of an intermediate size, somewhere in between the moon and Mars. The enormous amount of energy released in that titanic impact could have propelled a vast quantity of material toward space that ended up solidifying into the present-day moon. According to the defenders of this theory, most of the material hurled into space, and from which the moon was formed, came from the very body that struck Earth, and that celestial body was probably much poorer in iron than our own planet, which would explain Earth's lower density.

If, as most scientists dedicated to this question think, the existence of the moon owes itself to an accidental event, that being a planetary collision, we must congratulate ourselves for our good fortune. If such an event had not occurred, if the moon did not exist, the conditions of our beloved Earth would be quite different from the ones we know. In the first place, the day/night cycle would not be twenty-four hours, but much shorter, about fifteen hours. This is because the presence of a celestial body as large as the moon, positioned so close to us, slows the velocity of our planet's rotation owing to the friction of the tides. But the most dramatic influence brought to bear by the moon on our planet is that of stabilizing the variations in the angle of the axis of rotation of our world. This angle determines the cycle of the seasons, and its variations cyclically define the seasons every so many thousands of years. By stabilizing variations in *obliqueness*, or angle, of Earth's axis of rotation, the moon prevents sharp and wide-ranging fluctuations from occurring, acting as a ballast for Earth's climate. To put it more simply, if the moon did not exist, Earth would have been subjected, throughout its history, to very rapid and radical fluctuations in climate, which would have presented a serious obstacle to the development of life. If it were not for the moon being up there in the sky, there certainly would be no humanity down here to miss its presence.

In any case, the most popular hypothesis on lunar origin held today considers the moon either contemporaneous or subsequent shortly thereafter to the formation of Earth, its oldest rocks being a good indicator of terrestrial antiquity. With this idea in mind, the astronauts of the different Apollo missions were given the assignment of bringing back lunar rock samples to establish its age. Then a real scientific *thriller* began. One after another, the different expeditions were bringing back selenites whose age, to the consternation of the scientists, was between 3.1 and 4.2 billion years old. Finally, in the course of the last lunar mission, in December 1972, the astronauts of *Apollo* 17, Eugene Cernan and Harrison Schmitt, picked up a sample whose antiquity turned out to be at least 4.6 billion years old. Estimates of Earth's age were confirmed.

These results were later reinforced by the discovery of zircon crystals on our planet (found in more recent sedimentary rocks) that were 4.27 billion years old. In light of this, our planet's age seems firmly established. However, what really interests us is knowing which are the oldest sedimentary rocks, because only in their core will we be able to search for the remains of the first living things.

One of the places in the world where the oldest rocks on our planet can be found is in Isua, the western region of Greenland. There are strata there dating back around 3.8 billion years. The Isua rocks were formed in an atmosphere where water existed in a liquid state, where there were tropical temperatures and the phenomena of volcanism. Inside them, microspheres of organic matter have been found that appear to be fossils of primitive cells. This interpretation has little credibility today, but there is one aspect of the chemical composition of the Isua rocks that is really intriguing. It has to do with the ratio between the two stable isotopes of carbon.

The most abundant carbon isotope on our planet is Carbon-12 (o^{12}C), which accounts for slightly less than 99 percent of the total. The rarest but best known isotope is radioactive Carbon-14 (^{14}C), which is used to carry out radiometric dating. The least known isotope is Carbon-13 (^{13}C), which constitutes slightly less than 1 percent of terrestrial carbon. The difference among the three lies in the number of neutrons in their nuclei: the ^{12}C has six neutrons, and the isotopes ^{13}C and ^{14}C have seven and eight neutrons in their nuclei, respectively. The ^{12}C and ^{13}C are stable isotopes; that is, they do not experience radioactive disintegration.

What interests us is that living entities are quite discerning at the moment they choose the carbon dioxide molecules (CO_2) with which they will form their own organic matter, showing a slight affinity for those that include the isotope ^{12}C, rather than those comprised of ^{13}C. This preference for the light isotope produces a slight disproportion in their cells, in favor of ^{12}C, with respect to the isotopic ratio present on our planet.

Nevertheless, these paleo-biological-chemical proofs are hardly compatible with other types of evidence. The studies carried out on the moon indicate that our planet (and therefore, Earth too) was subjected to an intense hail of meteorites until some 3.9 billion years ago. It is plausible that some of them were sufficiently large so that the impact in question could have evaporated most of the water of the oceans. Although it is not completely impossible for life to originate under those circumstances (as we will discuss later), there is no doubt that until 3.9 billion years ago, Earth did not experience its best period for generating life. Then, if the Isua data are accepted as evidence of the existence of organisms 3.8 billion years ago, the lapse in time in which life was generated is reduced to less than 100 million years. Too little time, in the opinion of many, for the processes that generate life to have been able to take place.

Before continuing, it is appropriate to consider a point of great interest: 100 million years, is that a short or long period of time? In relative terms, that only represents a little more than two percent of the age of our planet. But this criterion is deceptive. When life appeared, 3.8 billion years ago (if in fact it took place in the Isua period), 100 million years represented, at most, 14 percent of Earth's age. Let's look, then, for another way to measure what percentage 100 million years represent. The ideal approach is to correlate this figure to the problem that is of interest to us.

It seems reasonable to admit that there is a much greater gap, in terms of complexity, between inert matter (including any combination of molecules) and a simple microbe, than between the latter and most complex of living organisms. However, judging by the fossil record data since the first living entity appeared (which would be very much like a microbe) until the appearance of the first *eukaryote* cells (the type of cells of which animal life is composed, for example), almost 1.7 billion years had to have elapsed; seventeen times more than the 100 million years that it took for the planet to generate life (if, as we repeat, the Isua evidence is accepted).

Nevertheless, many researchers attribute the long *delay* in the appearance of the eukaryote cells to the fact that a sufficient quantity of oxygen was not dispersed into Earth's atmosphere to support the chemical reactions on which life is based. According to this hypothesis, moving from the procaryotic stage to the eukaryote cell was held back by a low level of oxygen in the atmosphere and not because evolution was so temperate in its work. However, as we will soon see, the rate at which the increment of oxygen in question was produced is unknown, and thus we cannot know if eukaryotic organisms appeared shortly or long after the corresponding requirements for life were reached. The only thing that seems certain is that it took 1.7 billion years to do it. Judging by the speed at which biological evolution seems to act, 100 million years is a flash.

However, the opinion that 100 million years is a brief period for the development of the processes that generated life is based on the assumption that those processes, the majority being chemical in nature, acted with the same speed as biological evolution. The problem takes on a different perspective if, as two of the most eminent scientists in this field contend, Antonio Lazcano and Miller, this premise is not accepted in advance. According to these researchers, the *chemical evolution* that produced the first living organism from isolated biomolecules was produced at a much faster pace than the subsequent biological evolution, with a time lapse ranging from tens of millions of years to tens of thousands of years (in the most optimistic estimates) for the process to take hold. The tremendous authority of Lazcano and Miller on these subjects induces us to take their arguments very seriously.

On the other hand, the discovery in the 1970s of the ecosystems established around fumaroles, in oceanic abysses, allows us to contemplate the problem from another perspective. Existent in these places are some special bacteria that use reduced sulfur compounds, which are generated in these small underwater volcanoes, as a source of energy for producing their own organic matter, in a way similar to the one used by photosynthetic organisms. The tremendous resistance to the extreme heat existent in their habitats, with temperatures approaching 100 °C, has made it possible for *thermophiles* (heat-loving organisms) to dominate these other bacteria. Thermophile bacteria have also been found in places above water, like the hydrothermal springs of Yellowstone National Park. Thermophile bacteria are not only heat-loving, but they detest oxygen and only live in habitats devoid of it. On the other hand, the structure of one of their nucleic acids (the *ribosomic* RNA) is such that it is included in some of these thermophile bacteria in the bacteria group considered to be the most primitive (the so-called *Archaeabacteria*).

Maybe the organic matter in the Isua rocks, with their isotopic disproportion, was formed by organisms similar to current thermophiles, whose ancestors could have been generated a very long time ago in the great oceanic depths (in proximity to the underwater fumaroles), safe from the bombardment of meteorites. It is possible, but not all specialists are in agreement.

• • •

In the northwestern region of Australia lies a territory that, because of its extreme aridness, is known as the North Pole. There one finds the geological formation Warrawoona, constituted mostly of lava expelled in the core of shallow waters. There are also sedimentary rocks that are practically 3.5 billion years old, which include some very peculiar structures, called *stromatolites,* formed by the superimposition of hundreds of fine layers of sediments rich in calcic carbonate.

To the southwest of the Warrawoona region lies Shark Bay, whose depths appear covered with stromatolites similar to the ones from Warrawoona. But the Shark Bay stromatolites present a notable difference with respect to those of Warrawoona: they are still producing, and so it has been possible to discover the cause of their formation. The Shark Bay stromatolites are the result of the biological activity of communities of bacteria that form a layer a few centimeters thick on the bottom of the bay. In the opinion of many scientists, the Warrawoona stromatolites were also formed because of the activity of bacteria. But if this is so, what can we learn from those microorganisms that lived in this region of the planet 3.5 billion years ago?

There are other places in the world, among them the mouth of the Ebro River in Spain, where microbial mantles can be found like those of Shark Bay. The study of these microbial mantles has shown that they are comprised of several types of bacteria, which feed themselves in very diverse ways. Some, the *cyanobacteria* (until a short time ago known as cyanophycean or blue algae), produce the same type of photosynthesis as plants: they use water as the primary material and release oxygen as a waste product. There are other kinds of bacteria (red bacteria and green bacteria) that produce a different form of photosynthesis that neither uses water nor produces, oxygen but uses *hydrogen sulfide* (H_2S) as its primary material and releases sulfur. There are also bacteria that simply feed off the remains of dead photosynthesized bacteria. In fact, the microbial mantles that generate our present-day stromatolites are authentic miniature ecosystems.

But there is one aspect that determines life in these diminutive worlds: the oxygen released by cyanobacteria is toxic for the other types of bacteria. This provides cyanobacteria with a certain advantage. On asphyxiating their competitors, they eliminate them from the most suitable zone for carrying out photosynthesis: the highly illuminated upper layer of the microbial mantle.

However, microscopic structures have been found in the Warrawoona stromatolites that have been attributed to fossil cyanobacteria. Chemical proof has also been found indicating the presence of microorganisms that performed the same type of photosynthesis as green and red bacteria (the ones that release sulfur instead of oxygen). Thus, the Warrawoona stromatolites could be formed through the action of communities of bacteria similar to those that generate present-day stromatolites.

Even though there is no complete agreement, for many researchers the Warrawoona stromatolites and microfossils, and not the Isua rocks, constitute the oldest sure evidence of life on our planet. On the other hand, fossil stromatolites are known in other regions of the world, whose antiquity is very close to that of Warrawoona's. This is the case, for example, of the findings in the Fig Tree Formation of the Transvaal region (South Africa), whose age is somewhere around 3.4 billion years old. Following that, some 2.9 billion years ago, stromatolites became relatively abundant on the planet.

However, it is almost certain that the microorganisms that formed the Warrawoona stromatolites were not the planet's first living entities, but rather their long-evolved descendants. Photosynthesis, in whatever version, is an extraordinarily complex and refined process, which the first living organisms must not have experienced. Since the data indicate that the Warrawoona organisms were already using different types of photosynthesis (the one that produces oxygen and the one that generates sulfur) to synthesize their organic matter, they cannot be considered the first living entities to have appeared on our planet. The real pioneers, possibly bacteria of the thermophile type, must have lived before the Warrawoona period.

Even though they were not the first organisms on the face of Earth, we must not underestimate cyanobacteria, since they were responsible for the greatest environmental alteration in the planet's history. That was a fortunate event for us, because it consisted of the accumulation of the gigantic quantity of oxygen that our world's current atmosphere exhibits.

We have already said that our planet *should not* have unobstructed oxygen in its atmosphere, or at least not in its present quantity. First, there are no abiotic mechanisms for producing oxygen in great quantities. It is not a gas that is released during volcanic eruptions, and very small quantities of oxygen are generated in the upper layers of the atmosphere by the action of ultraviolet rays on water molecules. But in addition, oxygen is very reactive with a great number of chemical elements, therefore any oxygen molecule that is released is quickly captured to form sulfates and iron oxides *especially*. In fact, only 5 percent of the oxygen that is currently produced through photosynthesis is free in the atmosphere, the rest is trapped and converted, as we have already said, to sulfates and iron oxides.

Apart from the presumed fossils of Warrawoona cyanobacteria, which not all researchers accept as such, the oldest clear evidence of biological production of oxygen was established in the 2.7 billion-year-old stromatolites of the Tumbiana Formation in Western Australia. There, chemical proof of photosynthetic activity was found in environments very poor in hydrogen sulfide, and therefore, we can only be dealing with photosynthesis induced by water, the generator of oxygen. At that moment, Earth's atmosphere lacked free oxygen. For hundreds of millions of years, oxygen released by the action of billions of miniscule microbes was trapped by iron molecules dissolved in seawater, where it remained buried in the sediments in the form of iron oxides. But 2.3 billion years ago, after a heartbeat of 400 million years, microorganisms began doubling the circumference of the planet: oxygen obviously began accumulating in the atmosphere, to the point of affecting the composition of sedimentary rocks.

Pyrites and *uraninite* are two minerals that do not bear, without decomposing, prolonged exposure (we are talking about centuries of exposure) to an atmosphere with appreciable levels of oxygen. However, these minerals are not found in sedimentary rocks whose process of formation implies long exposure to the action of the atmosphere, less than 2.3 billion years old, but they are indeed present in sedimentary rocks formed before that date. The conclusion is clear: until 2.3 billion years ago, the amount of oxygen in the terrestrial atmosphere was so low that it was not capable of altering the pyrites and uraninites that proceeded to form part of the sedimentary rocks. Then, 2.3 billion years ago, the amount of oxygen in the terrestrial atmosphere increased to a level that prevented those minerals from remaining in the sedimentary rocks. It is calculated that, at that time, the amount of oxygen in Earth's atmosphere had reached a value equivalent to the hundredth part of what it is today.

The amount of iron oxide present in sedimentary rocks noticeably decreases after 1.7 billion years ago, indicating that the production of oxygen had already accounted for most of the iron dissolved in the oceanic waters. From that moment on, the accumulation of oxygen in the atmosphere accelerated, reaching current-day levels between 600 and 400 million years ago.

In this way, with their feat on a planetary scale, cyanobacteria determined the history of life in our world. Oxygen is a toxic element for living entities, and only those that were able to find mechanisms for tolerating it and using it in their own chemical reactions could spread throughout the planet. As an additional advantage, organisms capable of using oxygen acquire a greater amount of energy than those that do not use it. That surplus of energy is indispensable to the ability of multicellular organisms to maintain their complex organization. Without oxygen, we would not be able to live in this world.

But moreover, the increase in the level of oxygen in the atmosphere had a collateral affect of extraordinary importance for the development of life on our planet (especially for populating the emerging land masses): the appearance of the ozone layer. Ozone (O_3) forms in the upper layers of the atmosphere by the action of ultraviolet rays on oxygen molecules (O_2). The ozone atoms form a tenuous layer in the upper region of the atmosphere that filters the ultraviolet radiation that reaches Earth's surface. This radiation is lethal for living things, which, in the absence of the ozone layer, can only live under the protective mantle of oceanic waters.

In this way, the vast majority of organisms that populate our blue planet have incurred a debt beyond measure with the humble cyanobacteria. Paraphrasing Churchill, we could firmly conclude that never have the biggest owed so much to the smallest.

CHAPTER V

THE DEVOURERS OF LIGHT

THE FIVE FACES OF LIFE

There's an old Bob Dylan song whose verse (*"Man gave names to all the animals in the beginning, long time ago"*) perfectly reflects one of the most important jobs natural scientists have: the naming of living things. In reality, the goal of this task is none other than to establish the inventory of biodiversity; that is to say, to know how many different types of organisms there are on our planet. This is a question that allows for multiple responses depending on the meaning we give to the expression *different types of organisms*. If we are referring to the currently different known species, the answer is that their number greatly surpasses a million and a half.

All those organisms that are able to reproduce under natural conditions, and whose descendents may also be fertile, belong to the same species. Horses and mules are two different species because although they can be crossbred, the result, the she-mule, will be sterile. Distinguishing the different species is not a very complicated job in the majority of cases, and therefore, each species that is recognized and named by scientists also has a name customarily endemic to the different vernaculars (even though the corresponding one is not always exact).

However, not all species are equally distinct among themselves, but rather, there are groups of species that very much resemble one another. All felines are very similar, and we can say the same for beetles or pines. Using those common characteristics among species as a basis, scientists group them into more general categories each time, always as a function of their similarities. In this way, similar species are grouped into *genera*, and *genera* into *families*, which in turn are grouped into *orders*, which compose *classes*, which are grouped into *strains* (from the Latin *filum*, thread), and then finally, these are integrated into *kingdoms*.

Tigers, for example, are a species whose scientific name is *Panthera tigris* and belong to the genus *Panthera*, along with lions, jaguars, snow leopards, and cougars. In turn, this genus is integrated into the family *Felidae* (or *felids*) next to cats, lynxes, cheetahs, pumas, etc. The family of felids is included in the *Carnivora* order, to which the families of bears (*Ursidae*), hyenas (*Hyaenidae*), dogs and the like (*Canidae*), among others belong. The *Carnivora* order belongs to those that compose the mammalian class, which belongs to the Chordata strain (formed in great part by vertebrates) that, in turn, forms part of the animal kingdom.

So that, as you can see, we can formulate the following question: How many different types of living beings are there? At different levels, we can ask: How many different species are there in a specific family? or how many families in an order? or how many strains in a kingdom? or how many kingdoms are there in the entire group of organisms? This last question is, perhaps, the most interesting in that it can be reduced to something like this: How many *really* different types of living beings are there?

Perhaps you learned in school that only two kingdoms exist (that is to say, basically two different forms of living organisms): animal and vegetable. If this is correct, then we ought to begin by saying that today all biologists agree this is not so. Maybe it is the only question where there is a general consensus. Different specialists defend different classifications based on differing numbers of kingdoms, but what no one discusses is the fact that the diversity of living beings, as far as we know today, cannot be reduced to just two kingdoms.

How many kingdoms are there then? As we have already seen, there are different points of view, but the one that perhaps enjoys the most support, and is the easiest to understand, is that which holds that living things can be grouped into five distinct kingdoms: *monera,* which includes bacteria and cyanobacteria; *protista,* comprised of protozoans, unicellular algae, etc.; *fungi,* consisting of yeasts, molds, and mushrooms; *metaphyte,* or plants; and *metazoa* (metazoans), made up of animals.

What are the characteristics that distinguish the components of each kingdom? Or, to put it another way, what does it mean to be monera, protista, fungi, vegetable, or animal? The first distinctive feature, which allows us to divide living things into two large groups, is the level of organization: organisms can be made up of a single autonomous cell or formed from a group of cells functionally related to each other, so that when separated, they are not self-sufficient. The first group is designated as *unicellular* and the second as *multicellular*. Obviously, a great difference exists between the level of unicellular organization and multicellular, which justifies having them separately grouped. Of the five kingdoms, two are unicellular, monera and protista, and three are multicellular, fungi, plants, and animals.

Another major difference between living organisms has to do with the structure of the cells themselves. There are cells that are very simple and lack internal compartments (*organelles*), where the different functions can be independently generated. Not even genetic material is separated from the rest of the cellular components. This type of simpler, more primitive cell receives the designation of *prokaryotic*. There is another more complex type of cellular organization that is called *eukaryotic*, in which the cellular interior is compartmentalized into vesicles (organelles), where different functions are carried out. One of these vesicles separates the genetic material from the rest of the cellular interior and is called the *nucleus*. In short, prokaryotic cells lack organelles, including the nucleus, while eukaryotic cells clearly have organelles, among them the nucleus. The big structural and functional difference that exists between both cellular types is made clear if we keep in mind that there is not any multicellular organism whose cells are prokaryotic; in other words, the type of cellular structure limits the level of organization.

You have probably already deduced that the three multicellular kingdoms (fungi, plant, and animal) are composed of eukaryotic organisms (which is to say, eukaryotic-type cells). Of the two kingdoms of unicellular organisms, the monera presents prokaryotic organization, and the protista, eukaryotic.

Having arrived at this point, we are now familiar with the characteristics that define the kingdoms monera (prokaryotic unicellular organisms) and protista (eukaryotic unicellular organisms), while the fungi, plant, and animal kingdoms share the same type of cell (eukaryotic), as well as the same level of organization (multicellular). How are they distinguished from one other? The key lies in the different way of acquiring the support that the components of each of the multicellular kingdoms have.

Organisms that, like us, are incapable of autonomously generating their own organic matter, but rather use that which is produced by other living organisms, are called *heterotrophs*. Those other organisms that are able to synthesize their own organic matter from simpler molecules are called *autotrophs*. Within the kingdoms monera and protista, there are as many autotrophs as there are heterotrophs. Even species that can use both methods for their existence are found, depending on the environmental conditions. But within the kingdoms of multicellular organisms, there is much more discipline when it comes *to earning one's keep*: plants are autotrophs, while fungi and animals are heterotrophs.

Thus, plants are characterized as multicellular organisms, eukaryotes and autotrophs, while fungi and animals are distinguished as multicellular, eukaryotes and heterotrophs. The difference between fungi and animals resides in the different way that they behave as heterotrophs. Diverse forms of autotroph behavior also exist, and we will take up the subject in this chapter, leaving heterotrophs for later.

Most autotroph organisms gather their energy from solar light, through a process that is one of the great marvels of life: *photosynthesis*. This is a miracle that allows inorganic matter to be converted into organic matter. But before explaining what it consists of, and in order to be able to understand the far-reaching importance of that wondrous process, it is necessary for us to turn our attention to chemistry once more.

THE ECONOMY OF MOLECULES

In the third chapter of this book, we saw that the chemical energy resident in fuel is that which is responsible for powering the car and lighting its headlights. At the time, we also said this chemical energy proceeded, in the final instance, from solar light, and we refer you to previous pages for a clarification of these points.

In the first place, we should remember that not all cars use the same type of fuel. Some use gasoline, and others use diesel fuel.

The difference between the two fuels is found in their chemical composition: the first is made of organic molecules composed of five to ten carbon atoms, while the molecules of the second type vary from eleven to eighteen carbon atoms in their composition. In other words, the molecules that compose gasoline are smaller than those that compose diesel fuel. Apart from this difference in size, gasoline and diesel fuel molecules share an essential property: they are only formed by carbon and hydrogen.

Not all gasolines are equal, nor do they all cost the same. The fundamental difference among them lies in a property expressed as the *octane level,* which measures a gasoline's propensity to ignite: when the octane level is increased, the propensity of a gasoline to ignite is diminished. Of course, the greater the inclination, the lower the fuel efficiency and the greater the damage to the engine. Increasing the octane level of gasoline requires a higher degree of processing (technically known as *reforming*), which increases its price. Therefore, gasoline with a higher octane level is of better quality and more expensive.

The chemical composition of the different gasolines varies quite a bit, since different components can produce the same octane level. However, to simplify matters, we can say that the main ingredient in high-octane gasoline is a molecule known as 2,2,4-tetramethylhexane (C_8H_{18}). This molecule is composed of eight carbon atoms and eighteen hydrogen atoms joined by simple covalent bonds.

The tetramethylhexane molecules are placed in contact with the atmospheric oxygen (O_2), in the car's motor, producing a chemical reaction known as combustion, which releases a great quantity of energy. As a result of that reaction, water (H_2O) and carbon dioxide (CO_2) molecules are expelled through the tailpipe. Let's stop for a moment to analyze what happened. If we carefully study the reaction, we will observe that all the carbons, which were joined to the other carbon and hydrogen molecules before, are now joined to the oxygen molecules, and all the hydrogen molecules, which were linked to the carbon molecules, have also joined the oxygen molecules[1]:

$$C_8H_{18} + O_2 \Rightarrow CO_2 + H_2O$$

In other words, and this is the really important thing, the reac tion produced in the core of the car's engine consisted of breaking down the tetramethylhexane molecule in order to link all its atoms to the oxygen atoms. All combustion consists precisely of this: combining the atoms of organic molecules with oxygen. Therefore, it is not possible to make paper, wood, or a match burn in the absence of oxygen because the principal element of the chemical reaction is missing. It is not surprising that combustion forms part of a wider group of chemical reactions known as *oxidation*.

Surely at this point, one is probably asking himself what relationship all of this has with living entities. The answer lies in the fact that we and all other living organisms acquire our energy in the same way as the car engine: by oxidizing organic matter. Let's take the case of glucose, for example; it is one of the main fuels in our organism. Glucose is oxidized inside our cells, using the oxygen we take in through respiration in that process, and transformed into water and carbon dioxide, which we expel through our lungs. In the course of this process, a great quantity of energy is released, which is the one we use to perform our functions and remain alive.

1. To simplify the explanation, we have presented the basic reaction. In fact, two tetramethylhexane molecules react with twenty-five oxygen to produce sixteen molecules of carbon dioxide molecules and eighteen of water.

If you are not convinced, compare the formula of oxidation of the tetramethylhexane with that of the oxidation of glucose ($C_6H_{12}O_6$)[2]:

$$C_6H_{12}O_6 + O_2 \Rightarrow CO_2 + H_2O$$

In other words, organisms and cars acquire energy in an analogous way: oxidizing organic matter. Nevertheless, a radical difference exists in the manner in which the process is carried to completion. It is clear that if our cells oxidize glucose in the same way that tetramethylhexane is oxidized in the car's motor, we would burn like torches. In living beings, the processes for acquiring energy are performed in a very controlled manner, extracting the energy a little at a time, through numerous intermediary chemical reactions, the same as in a nuclear plant where the energy of radioactive material is extracted under very controlled conditions.

Oxidation of tetramethylhexane as well as that of glucose are chemical reactions that release energy. These are called *exergonic reactions*. There are also reactions that need to absorb energy in order to be realized, and these are known as *endergonic reactions*. If, for example, we want once more to combine carbon dioxide and water molecules to produce a new batch of gasoline (tetramethylhexane), or glucose, we would need to supply the same amount of energy that is released when they are oxidized. In other words, to produce a liter of gasoline from the carbon dioxide and water that escape through the car's tailpipe, we would need the energy produced by oxidation (combustion) of a liter of gasoline. But that is only a theoretical calculation; in fact, we would need to use more than one liter of gasoline to acquire that amount of useful energy,

2. We have not presented the exact reaction in this case either. Each glucose molecule reacts with six oxygen molecules, producing six molecules of carbon dioxide and another six of water.

since a good part of the energy supplied would be lost in the form of heat and could not be used in the process (because of, as always, the *cursed* second principle of thermodynamics).

But why is energy released when organic matter is oxidized? Or, to put it in different terms, why is energy released when carbon and hydrogen break apart and link with oxygen? The answer lies in the fact that the covalent bonds that are formed between different molecules require a different quantity of energy to come together: there are bonds of greater energy than others. Thus, the bonds between carbons, or between a carbon and hydrogen molecule, require more energy for formation than the bonds between hydrogen and oxygen, or between carbon and oxygen. The energy absorbed in the formation of a bond remains latent in the said bond, so that when it breaks away, the energy accumulated in the bond is released into the medium. In this way, when a high-energy bond (carbon-carbon or carbon-hydrogen) is broken to produce new, lower energy bonds (carbon-oxygen or hydrogen-oxygen), a surplus of free energy remains that can be used to make the car run or so that our cells will carry out their functions (plus the portion that is inexorably converted into heat).

To clarify these ideas, let's again establish the analogy between energy and monetary value, substituting in the previous paragraph the concept of energy for that of money. In this way, we can say that the covalent bonds between the different atoms have different prices, and there are more expensive bonds than others. The money used in the formation of a bond remains invested in that particular bond, so when it is broken, we will recover the money that it cost us to establish it. Following through with the comparison, it can be said that the carbon-carbon and carbon-hydrogen bonds are more expensive than the carbon-oxygen and hydrogen-oxygen bonds. What happens in oxidation is that expensive bonds are broken in order to form cheaper bonds, with remaining funds left over that can be used for other purposes.

But while energy produced in this way, in the oxidation of gasoline, remains free in the car's motor, tremendously increasing its temperature, the same thing does not happen in the core of cells. In fact, if the energy produced by the oxidation of glucose was released into the cellular medium in an uncontrolled manner, the cell would burn up. We have already said that the oxidation of glucose is produced through numerous intermediary reactions, which are carefully regulated. In some of these steps, small quantities of energy are released, which are immediately put into the formation of another type of high-energy bond, in the core of a molecule about which we have previously spoken: adenosine triphosphate or ATP.

ATP is formed by the union of a smaller molecule, adenosine, and three molecules of phosphoric acid. The bonds that remain joined to these phosphoric acid molecules, and to the rest of the ATP molecule, are high-energy bonds. Every time a sufficient fraction of energy is released in the oxidation process of glucose, it is put into establishing a high-energy bond between a molecule of *adenosine diphosphate* (this is, an adenine joined to only two phosphoric acid molecules), or ADP, and a molecule of phosphoric acid, generating one molecule of ATP. Through this procedure, for every molecule of glucose that is oxidized in any of our cells, thirty-eight molecules of ATP are produced and remain at the disposition of the cell to be used where and when it is necessary. Each time the cell requires energy to carry out any process, it will use as many molecules of ATP as it needs: on breaking its bonds (again yielding ADP molecules and phosphoric acid), it will recuperate the energy that it invested in forming them. Even though it may seem quite complicated, this mechanism allows the energy acquired in the oxidation of glucose to be used to the maximum.

Before concluding this section, and in order to treat chemistry with the respect it is due, we should be specific about the real meaning of the term oxidation, and its opposite, *reduction*. Previously, we gave you to understand that an atom oxidizes when joined to oxygen, and it reduces when linked to hydrogen, but these are only half-truths. Strictly speaking, an atom oxidizes when it surrenders electrons and reduces when it gains them. In most cases, this gain, or loss, is only relative. In fact, neither the carbon nor hydrogen atom loses electrons on bonding with oxygen. In the section on the covalent bond (which, if you remember, consists of sharing electrons), we already explained that the majority of times, one of the atoms forcefully attracts shared electrons and pulls them toward its nucleus, moving them away from their molecular *partners*. In this case, the atom that manages to bring the electrons closer is reduced (because it *gains* them), while the atom whose electrons are moved away from its nucleus is oxidized (since it *loses* them).

Of all the elements, the greatest force in attracting electrons is fluoride; any atom that links up with fluoride will see how this element draws its electrons away, thus oxidizing it. Fluoride boasts the title of oxidation champion, being capable of oxidizing oxygen itself. Surprisingly, in an oxygen-fluoride union, it is oxygen that gets oxidized and fluoride reduced! However, there is very little fluoride on our planet, and the little that exists is firmly joined to other elements, and therefore is not available for oxidizing organic matter. The second element in its ability to attract electrons, the runner-up in oxidation, is oxygen (in a close contest with chloride), which, besides being quite abundant, is also found isolated and ready to oxidize organic matter. Therefore, the oxidant par excellence of organic matter in our world is oxygen. It is important to point out that the fifth element in the list of oxidants, behind nitrogen, is sulfur. Later, we will see the importance of not forgetting this property of sulfur.

At the other extreme, hydrogen is the element that most easily surrenders its electrons to its covalent bond companion; it is the champion of reduction. This explains why when carbon is joined to hydrogen, it is reduced (because hydrogen surrenders its electrons to it), and why it is oxidized when it is joined to oxygen (because oxygen draws its electrons away from it). Since oxygen *exerts* more force than the carbon of the hydrogen electrons, hydrogen is oxidized even more upon exchanging carbon for oxygen.

If this electronic gibberish is very complicated to you, it will suffice to remember that when carbon combines with hydrogen, it is reduced, and when it is combined with oxygen, it is oxidized. Don't forget either that when moving from reduced carbon (joined to hydrogen or any other carbon) to oxidized carbon (joined to oxygen), energy is released, while with the inverse reaction, going from oxidation to reduction, energy is absorbed.

But to finish the comparison between cars and living things, let's consider a major question. Cars (their poor drivers, really) get their fuel from gasoline stations, but what about living beings? What and where are our service stations?

The Green Fire

All the organisms of our world are descendants of the first living organism that appeared on the face of the planet. From that first spark, the fire of life continued gathering force until it was transformed into a colossal flame. Today it burns everywhere: in the dark ocean depths, on the highest mountain peaks, in the smallest fissures of the Antarctic ice cap, at the bottom of caves, and in the cracks of rocks located dozens of kilometers above Earth's crust. Even though life has grown enormously diversified since its beginning, it has not reinvented itself. As living organisms, we are only transmitters of life from one generation to the next, but we do not generate it *ex novo*. In a certain and very profound sense, as living entities, we all remain an extension of that first organism. We are different flames of the same fire.

It was the Russian geologist Vladimir Ivanovich Vernadsky (1863–1945) who coined the expression *green fire* regarding the phenomenon of life. What Vernadsky wanted to emphasize in such a beautiful metaphor is the fact that living things behave like a special type of energy, which in turn, takes its nourishment from some other form of energy in order to maintain its existence. As we have already seen in previous chapters, as living beings, we serve as energy transformers, degrading energy and producing heat in order to survive, in the same way that fire consumes fuel in order to keep on burning. Thus the term *fire*. The energy that nourishes us, the one that keeps us alive, emanates as a last resort from solar light, which some organisms are capable of converting into chemical energy, in the form of organic molecules. At the heart of this process, known as photosynthesis, a *green* molecule is found—chlorophyll.

Photosynthesis constitutes the process that sustains the vast majority of life in our world and constitutes an authentic physical and chemical miracle. So that you may have an idea of what photosynthesis represents, imagine what it would mean to possess a device that converts the gases that escape through the tailpipes of our cars back into gasoline. That is exactly what photosynthesizing organisms are able to do: invert the process of breathing. In other words, it uses the energy from solar light, reuniting carbon dioxide and water in order to produce oxygen and organic matter (beginning now, we will simplify things by substituting the term organic matter for one of the biomolecules: glucose).

By a pleasant coincidence, photosynthesis was discovered in the Age of Enlightenment thanks to the research of the Englishman Joseph Priestley (1733–1804) and the Dutchman Jan Ingenhousz (1730–1799). The first of the two, Priestley, was one of those extraordinary men engendered by the eighteenth century. An incorruptible freethinker, his heterodox opinions on religion and politics caused him numerous heartaches throughout his life. His religious beliefs prevented him from attaining a university professorship, moving him to found, in 1758, a private academy, which became the most prestigious in Great Britain. Also owing to his religious opinions, he was barred from Captain James Cook's (1706–1790) second expedition, which he was to join as an astronomer at first.

In the political arena, Priestley also swam upstream against the dominant ideas of his time. Thus, he was one of the few defenders, in England, of the causes of the American Revolution, first, and the French Revolution, afterward. On July 14, 1791 (on the second anniversary of the storming of the Bastille), his support of the latter movement dearly cost him when an infuriated mob destroyed his home and laboratory. Nevertheless, Priestley remained in England for three more years, but in 1794, he went into exile, across the ocean to the recently born American republic, where he was warmly received by his friends Thomas Jefferson (1743–1826), among others and John Adams (1735–1826). In the United States, he was offered the chair in chemistry at the University of Pennsylvania, but he turned it down in order to found, along with his wife, a colony of religious exiles from Great Britain. This project never came to fruition. When the conflict between Jefferson and Adams arose, the incorrigible Priestley sided with

Jefferson. Jefferson's election as the third president of the United States, in 1800, saw Priestley spend his final years, for the first time in his life, in rapport with those in power.

Throughout his life, Priestley took an interest in a wide range of subjects, such as grammar, politics, theology, education, and chemistry. He rose to the occasion in all of them, even though it was his scientific work that earned him his passage into posterity. It was Benjamin Franklin (1706–1790), whom he met in 1765, who drew Priestley's attention to the phenomenon of electricity. His first important success was none other than the discovery (in 1767) that graphite conducts electricity. He also correctly deduced, from Franklin's experiment, the inverse quadratic relationship that maintains the force of attraction between two electrical charges with the distance that separates them (even though this discovery went unnoticed at that time).

However, his interest soon shifted from nature and the properties of electricity to the problem of the chemical composition of air, where he made his most important discoveries. His first work in this field was connected to his failed participation in Captain Cook's second voyage. This expedition was planned with consummate care in order to avoid, to the greatest possible measure, deaths owing to diseases caused by deprivation or malnutrition. The precautions proved so effective that only one man, out of a crew of 112, died from disease on a trip that lasted three years. Within that framework, Priestley undertook solving the problem of keeping water potable for the longest period of time possible. With that end in mind, he conceived of a way to incorporate carbon dioxide into water, with the idea that this would prevent contamination. In effect, carbonated water (or soda), which is the basis for a great many of our beverages, was Priestley's invention.

Most of Priestley's biographies underscore his discovery of

oxygen (in 1774) as his most notable scientific achievement. However, the fatherhood of that discovery is debatable since, in fact, the first one to isolate the gas was the Swiss chemist Carl Wilhelm Scheele (1742–1786), even though the publication of his discovery appeared (for editorial reasons) after Priestley's work. In any case, both made the discovery independently. However, neither Scheele nor Priestley correctly interpreted their discovery, owing to the fact that both were partisans of the theory of *phlogiston*, very much in vogue in the eighteenth century.

Priestley and Scheele were convinced of the existence of a particular substance—phlogiston—that unquestionably took part in the processes of combustion and respiration. According to this theory, phlogiston was an elastic fluid that formed part of the composition of inflammable substances. When those substances burned, phlogiston was released and then combined with air. Once air was saturated with phlogiston (transforming itself into phlogisticated air), combustion ceased. Partisans of the theory of phlogiston thought, as happened in combustion, that the respiration of organisms released phlogiston into the air. In the same way that phlogisticated air (that is, air saturated with phlogiston) did not support combustion, neither did it allow respiration.

What Priestley really discovered is that when mercury oxide is heated inside a bell jar with *phlogisticated air* (which permits neither combustion nor respiration), a change is effected in the nature of that air, so that it is again possible for a flame to burn or an animal to breathe. Priestley interpreted the result to mean that the air had been *dephlogisticated;* in other words, freed from the phlogiston. In fact, he called the gas obtained in his experiment *dephlogisticated air*. The correct interpretation of Priestley's results was the work of Lavoisier, who rejected the existence of phlogiston and instead suggested the existence of a gas (oxygen) that combined with other substances in the processes of combustion and respiration. While Priestley thought he had simply freed the air in the bell jar from phlogiston (he had *dephlogisticated* it), Lavoisier correctly deduced that, upon being heated, the mercury oxide had released a new gas (oxygen) into the air inside the bell jar. The polemic between Lavoisier and Priestley did not end until their deaths.

Thus, even though Priestley and Scheele are deserving of recognition for having been the first to isolate oxygen, it was really Lavoisier who recognized it and established its existence as a chemical element. It is your decision as to who was the authentic discoverer of this gas.

In this context, there is an aspect of Priestley's research that especially interests us. This is an experiment that he carried out in searching for a way to acquire dephlogisticated air (or air rich in oxygen). Subsequent to having a candle burn inside a bell jar until the flame went out, Priestley observed that introducing a sprig of mint into the medium after several days was sufficient to make the flame burn again. With this experiment, Priestley established that green plants produce oxygen (dephlogisticated air, in his words), and without realizing it, he had taken the first step toward the discovery of photosynthesis.

But the physiology of plants was not the object of Priestley's interest, and so he did not pursue that line of research. The one who was really interested in how the vegetable kingdom functioned was the Dutch physician and botanist Jan Ingenhousz. In an admirable series of experiments (published in 1779), Ingenhousz corroborated Priestley's results and unquestionably established that light from the sun, and not its heat, was responsible for green plants producing oxygen (which Ingenhousz also called dephlogisticated air). A religious man, Ingenhousz thought that this characteristic of plants echoed the wisdom of the Creator, who remained vigilant in restoring the quality of air that became contaminated through animal and plant respiration, the latter occurring during the night.

Since plants stopped producing oxygen at night, but not their respiration, Ingenhousz sounded the alarm over the danger run by those persons who slept with plants in their rooms at night. In his opinion, plants depleted the air during the hours of darkness, placing the unsuspecting person at risk. This idea that plants are harmful at night constitutes the greatest success in the history of scientific dissemination, since today, more than two hundred years after its formulation, it persists as a very common topic, at least in Europe. This idea was so widespread on our continent that in the *New York Times* science section (September 12, 1995), a response was given to a reader who asked if his European friends were right in warning him of the danger of sleeping with plants in his room. The specialist's answer was that only those persons who sleep in a very small, poorly ventilated room filled with plants ran some risk. But he concluded by adding that another person sharing the room with the individual in question would consume much more oxygen than any of our normal household plants.

The light from which plants acquire the energy to produce photosynthesis is a complex phenomenon whose understanding stumped physicists for more than two hundred years. Today we know that it shares many of its properties with other physical phenomena that we list under the heading of *waves*. The ripples in the ocean are a good example of waves, and in observing them, we can come to know some of the basic properties of this type of phenomena. If you place yourself somewhere along the coast, with a chronometer to count the number of waves that pass over a point of reference in a fixed amount of time (let's say a minute), it will

Figure 11: *Lavoisier analyzing the atmospheric air (in* The Atmosphere: The Great Phenomena of Nature *by Camille Flammarion. Montaner and Simón, Barcelona, 1902).*

establish a fundamental parameter for any wave: its *frequency*. The more waves that pass in a fixed period of time, the greater will be the frequency of the swells. Another characteristic you can consider is the distance between the crests of the waves; that is to say, the *wavelength*. Evidently, the greater the particular wavelength (and therefore, the greater the separation between waves), the fewer the number of waves that will pass over the point of reference in each unit of time. Or, in other words, the frequency and wavelength are inversely proportional: the greater the one, the lesser the other.

Another phenomenon of routine observation and also of great interest is the fact that light from the sun, which is white in color, breaks down into different colors when it passes through raindrops. Thus, a rainbow is formed when light from the sun cuts across raindrops or splashes of water are launched into the air from a waterfall or geyser. The explanation for this beautiful phenomenon lies in the fact that light from the sun is made up of a beam of other lights that separate when they pass through a transparent body (because of the physical phenomenon known as *refraction*). Each one of those different lights that compose solar light corresponds to a different wave, with its frequency and distinct, characteristic wavelength. These lights can more properly be designated as electromagnetic radiations. From this point forward, we will use both terms, light and radiation, synonymously.

If we study the rainbow, we can distinguish the different colors (lights or radiations) that compose the white light: from red at one end, to violet at the other. Red is the one that has a greater wavelength, and therefore, it is also the one with the lowest frequency. On the other hand, the violet light presents the greatest frequency and is the one with the smaller wavelength. But the colors of the rainbow are only those that the human eye can perceive. In solar light, there are other radiations with an even lower frequency (and a greater wavelength) than the red light: it is ultraviolet light or ultraviolet radiation. There are also lights beyond the violet color that are imperceptible to us; in other words, lights with a greater frequency (or, if you prefer, a shorter wavelength): this is ultraviolet light or radiation. Beyond infrared radiation, one discovers radio waves with frequencies greater than that of ultraviolet light; these are X-rays.

Each one of these lights contains energy, and the higher the frequency, the greater the quantity. Therefore, ultraviolet rays are capable of burning our skin, because they hold a great quantity of energy. Surely, you probably remember that this is not the first time we have associated energy with light. In the third chapter, we already said that red light (a low-frequency light) was associated with high-energy photons. If you remember, photons are the equivalent of quanta: the minimum quantities of energy that matter is capable of absorbing and/or emitting. So that when plants capture energy from solar light in order to perform photosynthesis, they are, in fact, capturing photons, or minimum units of energy in the form of light.

Since not all lights are equal, each one having photons of different energy, we can ask ourselves whether plant life can use all of them to its advantage or whether, on the other hand, plants behave selectively when it comes to using one type of radiation or another. The fact is that plants only use visible light. Infrared radiation contains little energy, and ultraviolet light has too much. Merit lies in the middle, and that is the range of radiation that composes visible light. However, here we present you with a riddle: from among the lights that compose the visible spectrum, which one or which ones do plants prefer? The answer may seem astonishing to you: all except the green one. Still, if you stop for a moment to think about it, you will discover the answer is obvious. If the plants that perform photosynthesis are green, it is because they reflect precisely that type of light. They reflect it because they do not use it. If they were to use all the radiation that composes white light, plants would not reflect any of them and would be black in color.

The Great Miracle

In general terms, photosynthesis is the opposite process of respiration, and therefore, its general formula is:

$$CO_2 + H_2O + \text{light} \Rightarrow C_6H_{12}O_6 + O_2$$

It is easy to deduce that what happens in photosynthesis is that the molecules of water (H_2O) diffuse away from each other, and their hydrogen atoms end up joining the carbon atom of the carbon dioxide (CO_2) molecules, which, in turn, bond together. In the course of the process, the oxygen molecules remain free. However, this is a very rudimentary description of a process that is comprised of numerous intermediary steps. We should clarify this a little more. In the first place, it is necessary to specify that the oxygen that is released into the atmosphere comes from the water molecules alone and not from the carbon dioxide molecules.

In the second place, it is important to explain that the hydrogen atoms that travel from the water molecules to the carbon dioxide molecules do it in a very special way. When the bond between the hydrogen atoms and oxygen atom of a water molecule is broken, the hydrogen atoms are separated: on the one hand, its nucleus, and on the other, the electron that participated in the covalent bond. Thus, the fragmentation of the water molecule produces one atom of oxygen, two hydrogen nuclei, and two electrons. Since *normal* hydrogen atoms only have a single proton in their nucleus (remember that the hydrogen isotope called deuterium has, in addition, a neutron in the nucleus), we can say that for every fragmented water molecule, one oxygen atom, two electrons, and two protons (the two hydrogen nuclei) are acquired.

In other words, photosynthesis consists of the splitting of a water molecule and the orderly transfer of its hydrogen constituents (one electron and one proton for every atom) to the carbon dioxide carbons. This process takes place in two stages: in the first one, the water molecule splits, and the electrons and protons are carried to their corresponding receptor, where they are again joined together; in the second stage of photosynthesis, the hydrogen atoms are surrendered to the carbon of the carbon dioxide molecules so that they will come together and form organic molecules. The first phase is called the *luminous phase* (or *light reaction*), because it requires light for it to be carried to completion, and the second part is known as the *dark phase* (or *dark reaction*), because until a short time ago, it was thought that it was totally independent from light. In plants, both phases take place in a cellular organelle, called the *chloroplast*, where solar light is captured and all the chemical reactions of photosynthesis are produced.

However, to split the water molecule and remove two electrons from it is a chemical feat. If you remember, hydrogen is the element that has the greatest facility for surrendering its electrons, while oxygen is the second element as far as zeal for capturing electrons. Together, hydrogen and oxygen form an extraordinarily stable partnership, held together by low-energy covalent bonds. How is it possible to disrupt such a stable molecule and transfer two electrons from it to the *tyrannical* oxygen molecule? The difficulty in achieving it is so great that, photosynthesis aside, this is a process that is rarely produced in nature. The key to this feat is found in a complex group of molecules known as the *photosystem*.

Chloroplasts have two different photosystems that fulfill different missions and are known as *Photosystem I* and *Photosystem II*. Both are composed of a good number of proteins (many of which are enzymes that catalyze the different chemical reactions) and other molecules known as *pigments*.

Pigments are molecules that, because of their peculiar structure, are able to absorb a photon and elevate one of its electrons to a high energy level (do you remember the energy levels at which electrons position themselves, which we explained with the example of the stairway whose steps are set at different heights?). When it is found in that situation, with an electron promoted to a high energy level, it is said that the pigment molecule is *excited*.

The principle pigment in both photosystems is *chlorophyll*. The chlorophyll of Photosystem II is excited by the light whose wavelength is 680 nanometers (a nanometer is one billionth part of a meter), while the chlorophyll of Photosystem I captures the light of 700 nanometers of wavelength. Other pigments exist in both photosystems that are capable of absorbing the energy from lights of other wavelengths and channeling that energy to the chlorophyll molecules. This group of *accessory* pigments is called *collector antennae* (there is one per photosystem).

In fact, only one of the two photosystems participates in splitting the water molecule and freeing the two electrons from its hydrogen atoms. This is Photosystem II. The designation is disconcerting for students of biology, because Photosystem II is the one that initiates the process of photosynthesis, and intuitively, since it is the first to act, it seems more appropriate that it be designated Photosystem I. However, this terminology has its logic: Photosystem II receives this designation because it was the second of the two systems to appear in the line of evolution.

When the chlorophyll molecule of Photosystem II is excited by a photon, one of its electrons is transferred to a chain of molecules known as a *transporting chain of molecules*, which leads it to Photosystem I. The transfer of the electron by this transporting chain generates energy with which the cell is able to synthesize a molecule of ATP (from ADP and a molecule of phosphoric acid), which, as we have already seen, is the energy unit of living beings. In this way, the energy from the light that excited the chlorophyll molecule has gone into forming high energy bonds of ATP. Radiant energy has been transformed into chemical energy.

The chlorophyll of Photosystem II now finds itself minus one electron and *needs* to capture another one in order to recover chemical stability. Chlorophyll's eagerness for electrons unleashes the fragmentation of the water molecule in order to snatch away the electrons belonging to the hydrogen atoms (through a complex mechanism known as a *photolytic oxygen clock*). These electrons reach (one at a time) the chlorophyll, which will again lose them when absorbing energy from light (a photon for each electron), initiating a new cycle.

Let's return to the electron that we had left at the doors of Photosystem I, where it is captured by the corresponding chlorophyll molecule. The arrival of a photon excites the chlorophyll in Photosystem I and causes it to lose the electron that it had just accepted from Photosystem II. From this point forward, the electron can follow two courses of action. In the first instance, it can be led back to the electron transport chain, in order to advance the formation of ATP again. This mechanism is known as *electron cyclical flow*. The other alternative consists of being transported, via the inevitable chain of intermediary electrons, to its final acceptor, a molecule that bears the impressive name of *nicotinamide adenine dinucleotide phosphate,* but which is normally

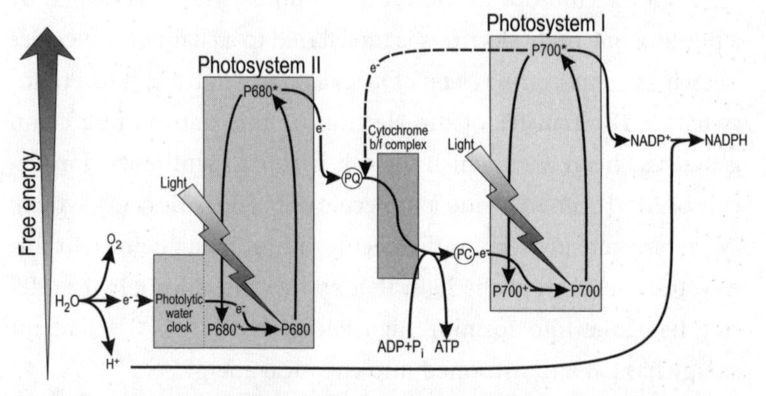

Figure 12: *A diagram of the light phase of photosynthesis. At the core of Photosystem II, a photon (Light) reaches the chlorophyll molecule (P680), exciting it. The excited chlorophyll (P680*) loses an electron (e^-) and is transformed into a molecule hungry for electrons (P680+). This situation induces the fragmentation of a water molecule through the photolytic water clock. For every broken water molecule, two electrons are released and subsequently captured by the photolytic clock, which supplies them to the chlorophyll one by one. The electron that was released in Photosystem II (e^-) travels to Photosystem I through an electron transport chain, composed of, among other molecules, plastoquinone (PQ), the cytochrome b/f complex and plastocyanin (PC). The transfer of the electron by this transport chain releases sufficient energy to generate ATP (adenosine triphosphate) from ADP (adenosine diphosphate) and phosphoric acid (P^i). Once in Photosystem I, the electron is captured by a molecule of chlorophyll with one less electron (P700+), which is then converted into stable chlorophyll (P700). When a photon (Light) arrives, this molecule of chlorophyll is excited (P700*) and loses the electron that it had captured, returning to a situation in which it is minus one electron (P700+). The electron can return to the transport chain (the discontinuous line in the diagram) in a process called the electron cyclical flow in order to advance once again the synthesis of ATP. The electron finally attains one molecule of $NADP^+$, which accepts it alongside one proton (H^+), incorporating one atom of hydrogen into its structure (NADPH).*

referred to by its initials: $NADP^+$ (the $^+$ symbol indicating that it is found in its oxidized state, and it is able to accept an electron). Upon reaching $NADP^+$, the electron again joins a proton to form a new hydrogen atom and become part of a NADPH molecule (the final H indicating that the molecule now has one additional hydrogen atom).

The luminous stage (light reaction) of photosynthesis ends here, over the course of which we attended, with the participation of light energy, the fragmentation of a water molecule, and its transformation into chemical energy, in the form of ATP and NADPH.

In the dark stage (dark reaction) of photosynthesis, those ATP and NADPH molecules are used to synthesize the corresponding organic molecules. The group of chemical reactions (more than a dozen), which comprises the process of glucose molecules synthesis from the carbon dioxide molecule, is known as the Calvin Cycle[3], in honor of its discoverer and Nobel Prize winner for chemistry, Melvin Calvin (1911–1997).

But the Calvin Cycle is not the only process that takes place in the dark reaction of photosynthesis. The products of the luminous reaction (ATP and NADPH) are also used to produce ammonia and hydrogen sulfide (essential for synthesizing the amino acids) from two salts present in soil and water: sulfates and nitrates.

Alternative Menus

Photosynthesis, such as we have described it, is not a process that is the exclusive patrimony of plants, for it is also carried out by cyanobacteria and algae. In fact, most of the oxygen produced throughout the history of our planet, and which continues being produced today, is released by those modest unicellular organisms.

On the other hand, even though this type of photosynthesis is a dazzling process, which generates the preponderance of organic matter, and therefore, life on our planet, it is not the only way that living beings have found, throughout history, to procure support for themselves on their own. To begin with, as we already stated in the chapter on the origin of life, there are

3. In fact, the Calvin Cycle produces glyceraldehyde 3-phosphate (three carbon atoms each), which can be used as a starting point for the synthesis of glucose, as well as that of other biomolecules like lipids and amino acids.

other different types of photosynthesis. These *other types* are confined to the world of bacteria and display notable differences from the *standard* photosynthesis of plants, algae, and cyanobacteria.

In the first place, they differ in the molecules to which electrons are drawn (or *electron donors*) thanks to radiant energy. If the photosynthesis of plants uses water as a source of electrons (and protons), the different alternative types of photosynthesis use a pleiad of other molecules, which includes different compounds of sulfur (like hydrogen sulfide: H_2S), molecular hydrogen (H_2), different organic molecules (like *succinate* and *malate*), and even iron (in its reduced form, ferrous iron). When water molecules do not split, the organisms that produce these other types of photosynthesis do not release oxygen into the atmosphere, and here we are speaking of *anoxygenic* photosynthesis (meaning, a nonproducer of oxygen), as opposed to the photosynthesis that releases oxygen and is termed *oxygenic.*

A good example of anoxygenic photosynthesis is that of red and green sulfur bacteria. When, on previous pages, we stopped to explain which chemical elements were good oxidizers (or in other words, electron acceptors or, if you prefer, hydrogen molecules), we mentioned sulfur among them. Just like oxygen, sulfur can form covalent bonds with two hydrogen atoms in order to form oxygen sulfide (H_2S). It is easy to see that this molecule is similar to the water molecule (H_2O). However, in this type of photosynthesis, light energy is used to break the bonds between the sulfur and hydrogen atoms, and electrons and protons from the released hydrogen atoms are used to generate chemical energy (ATP) and hydrogen transport molecules (NADPH), the same as in oxygenic photosynthesis. However, unlike the latter, elemental sulfur (S^o) and water are released rather than oxygen (O_2). Compare the general formulas of both processes, and you will be able to appreciate the similarities and differences:

Oxygenic photosynthesis:

$$CO_2 + H_2O + \text{light} \Rightarrow C_6H_{12}O_6 + O_2$$

Anoxygenic photosynthesis (sulfur bacteria):

$$CO_2 + H_2S + \text{light} \Rightarrow C_6H_{12}O_6 + S^O + H_2O$$

But the differences do not end here. The organisms that produce any type of anoxygenic photosynthesis are only equipped with one photosystem in their cellular machinery (and not two, as in the case of plants, algae, and cyanobacteria), whose chlorophyll is not exactly equal to that of the photosystems of plants either. To illustrate this, the red bacteria of sulfur only have one photosystem whose chlorophyll absorbs photons of light with a wavelength of 870 nanometers (appreciably greater than the 700 nanometers of the chlorophyll in Photosystem I). The bacterial photosystem is equivalent to Photosystem I and was the first to appear in evolution, although organisms (cyanobacteria) stored in both photosystems soon appeared. The appearance of Photosystem II allowed the use of water as an electron donor, which involved two great advantages. In the first place, water is a much more abundant *fuel* in our world than any other alternative electron donor and, moreover, is universally distributed. The second advantage was that, with oxygen being a stronger oxidizing element than sulfur, the splitting of the water molecule allows a greater quantity of chemical energy to be generated.

In our world, some microorganisms still remain as evidence of the time when Photosystem II appeared. Depending on environmental conditions, the cyanobacteria *Oscillatoria limnetica* can carry out either oxygenic photosynthesis, with both photosystems working in tandem, or anoxygenic photosynthesis, using only Photosystem I (in an analogous way to the bacterial photosystem).

Toward the end of the decade of the 1880s, the great Russian microbiologist Sergei Nicolaevitch Winogradsky (1856–1953) made a surprising discovery: there are microorganisms (bacteria) capable of autonomously generating their own organic matter in the absence of light. Up to that time, it was believed that all of Earth's living organisms rested their survival on the ability of photosynthetic organisms to convert solar light into organic matter. These autotrophic organisms that do not use photosynthesis are called *chemoautotrophs*.

Like photosynthesizing organisms, chemoautotrophs require an electron donor and a source of energy to convert carbon dioxide to organic matter. The electron donor can be quite varied: hydrogen bacteria use that very element (H_2); colorless sulfur bacteria use hydrogen sulfide (H_2S); nitric bacteria use ammonia (NH_4), and iron bacteria are used in the ferrous form of this element. As a source of energy, these organisms use the type of bacteria that are released when the electron donors are oxidized. Thus, in the presence of oxygen, hydrogen is oxidized to water, sulfur to sulfide, ammonia to nitrites, and ferrous iron is transformed into ferric bacteria. In these processes, chemoautotrophs acquire, at the same time, the electrons and energy they need to synthesize their organic matter.

Surely, you have probably noticed that they all need the presence of free oxygen in order to produce the oxidizing reactions with which they obtain their energy. Since free oxygen is a product of oxygenic photosynthesis, we can conclude that these organisms also depend on that process and, therefore, solar light. This is true. But it is also true that there are chemoautotrophic organisms that do not use oxygen to oxidize the electron donors, but rather other oxidizers, weaker ones, like carbon dioxide, sulfur, sulfates, nitrites, or ferrous iron. These chemoautotrophs that do not use oxygen are not, in effect, dependent on solar light or any sort of photosynthesis. Today there are many specialists who believe that the first forms of life to appear on our planet were organisms of this classification.

Be that as it may, as we have already seen, oxygenic photosynthesis is the process responsible for the formation of most of our world's organic matter, as much now as in the distant past. Since our fossil fuels—carbon, gas, and petroleum derivatives—were formed from the remains of billions of former organisms whose organic matter was synthesized through radiant energy, don't forget, the next time you stop to fill the gas tank in your car that, when it comes down to it, the fuel dispensed at gas stations is nothing more than solar light . . . transformed into chemical energy.

Chapter VI

The Indispensable Fungi

War and Peace

For many persons, the terms fungus and mushroom are synonymous. But this is a mistake; the mushroom is a structure of spores that germinate and form some type of fungi at the moment of reproduction.

But if the fungus is not the mushroom, then what are fungi like? It is a little difficult to imagine, since fungi give no evidence of tissues or authentic organs. In fact, their cells are not even completely separated from each other, as in the case of plants and animal life. Fungal cells form tiers, called *hyphae,* interconnected in a reticular group called the *mycelium.* The mycelium of a simple fungus looks very much like a cotton swab.

Lacking organs, fungi do not have a digestive tract nor, as a consequence, a mouth. Therefore, fungi do not take in food that is to be internally digested, but rather, each cell, acting on its own, releases substances that decompose organic matter into its primary molecules and then absorb them.

Most fungi obtain their organic molecules by decomposing dead parts of other living organisms, like fallen leaves, tree bark, or the epidermis of terrestrial vertebrates. This is an essential chore that allows organic matter to be recycled into the ecosystems, where fungi and bacteria are the main protagonists. The name for this docile method of gathering sustenance is *saprophytism*.

Besides saprophytism, many fungi acquire their organic molecules through a much less innocent procedure: *parasitism*. Fungi that use this method for earning their keep grow on the host organism or, more frequently, inside its tissues. In this latter case, extensions, called *haustoria*, which sprout from the fungus's cells are introduced into the parasitic cells and absorb nutrients from them. Evidently, this form of cellular vampirism is detrimental to the cells that experience it, since it results in their death.

Parasitic fungi in humans are scarce, but some of them can be terribly troublesome (like the one that produces athlete's foot, *Trychophyton interdigitale*, or the cause of ringworm, *Microsporum audoninii*) or even fatal. Among these latter cases is found the species *Candida albicans*, which frequently lives in the mouth, digestive tract, and vagina of humans. In those instances, this particular species is kept under control by the competition sustained by other microorganisms of the bacterial flora. But when the balance maintained by this flora breaks down, the fungus can proliferate and cause pathologies, like aphtha in infants, vaginitis, or underarm and groin skin infections. If their spores are inhaled and the fungus invades the lungs, serious illness, and even death, can result if proper treatment is not given.

Insects suffer many more infections from fungi than vertebrates. This is a fortunate circumstance for us, because parasitical species can be used in controlling plagues. For example, in the struggle against populations of the *Anopheles* genus of mosquitoes, propagators of malaria, fungi of the *Caelomyces* genus are used to destroy the larvae of those harmful insects.

Fungal parasites that affect plants constitute an authentic

plague, and it is calculated that they are responsible for about 80 percent of the total number of diseases suffered by crops. The fungus known as ergot of rye (*Claviceps purpurea*), which infects that particular grain, produces a highly toxic substance that created havoc during the Middle Ages among unsuspecting populations of people that consumed this cereal. Another example of the devastating consequences of parasitical fungi in crops happened in 1840 in Ireland, when a potato parasite (*Phytophtora infestans*) ruined the potato crop and caused a tremendous famine that resulted in thousands of deaths (there are those who place the estimate at one million dead) and forced a good portion of the population to emigrate.

Not even fungi themselves are safe from their pals, and there are mycoparasitical species capable of infecting other fungi and living at their expense. Even some of them, as is the case with *Penicillum vermiculatum*, can cause death in infected fungi.

However, the death of the host is never a good business for the parasite, because their own survival depends on that of its victim. In light of this fact, some fungi, over many millions of years of evolution, have found ways of improving their relations with the organisms they invade. The trick consists of giving something in exchange for the parasitism. Those fungi, which, over the course of thousands of generations, have come to favor their victims in some way, are assured of a better supply of food, augmenting their possibilities of survival in the process. Two situations have emerged in this way and perfected themselves, whereby the fungi benefit the organisms they invade: the formation of lichens (or *lichenization*) and *mycorrhizas*.

Lichens are living organisms that grow on barren soil, the surface of rocks, or other organisms (tree bark, for example). They can be scaly, leafy, or vinelike in appearance. It was believed for a long time that lichens were a type of organism similar to plants. However, that is not the case. Lichens are formed by the joining of two different living organisms that thrive in intimate association: a fungus and an organism capable of carrying out photosynthesis (a unicellular alga or cyanobacteria). The fungus that forms part of a lichen is classified as a *mycobiont*, while the photosynthesizing organism is called a *photobiant*. Even though there are numerous cases in which one type of specific fungus inevitably selects the same type of alga (or cyanobacteria), it also frequently happens that there can be various mycobionts and photobionts in a single lichen. But this kind of cellular promiscuity does not always possess the tolerance of the mycobiont. Sometimes, a particular fungus infects a previously formed lichen, and far from proving agreeable to the patron mycobiont, it stockpiles the photobiont's hospitality until it makes its competitor disappear. Botanists have christened this nasty form of behavior *cleptobiosis*.

Problems of pairs aside, what is true is that together, mycobiont and photobiont can live in extreme environments, where they would not be able survive separately. In fact, lichens extend from the arctic to the antarctic regions and can also prosper in high mountain and desert climates. Among the few environments hostile to lichens are areas where there is air pollution, like in large urban centers. Given their extraordinary sensitivity to atmospheric pollution, lichens constitute a magnificent bioindicator of air quality.

At the heart of a lichen, the fungal hyphae protect the algae cells from excessive light and aridity. Moreover, fungal cells are able to capture water from environmental humidity and surrender it to the algae. They also carry diverse nutrients that their cells are able to absorb from the substratum. But nothing is free in nature. The fungus does not go to great pains for the well-being of the algae out

Figure 13: Above: The internal structure of a simple lichen. The hyphae of the fungus (mycobiont) completely envelop the cells of the photobiont (f). Below: The Xanthoria lichen with the reproductive structures (apotecia) of the fungus.

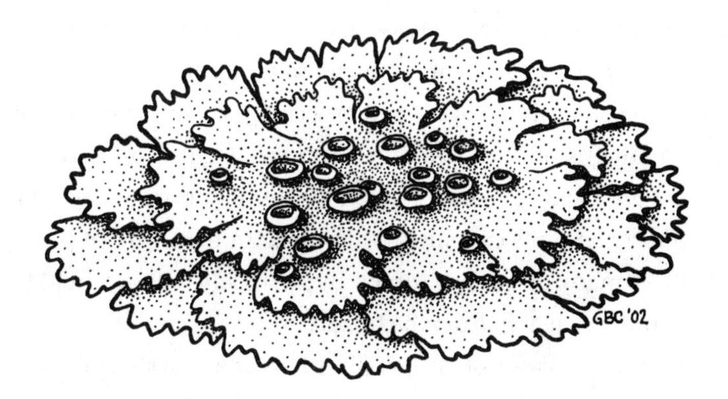

of any altruistic motive. Haustoria leave their fungal cells and penetrate the alga's cells, where they absorb the organic molecules that the alga synthesizes through photosynthesis.

It is a marriage of convenience between the fungus and alga, because if the environmental medium is favorable to the growth of either of the two members (or both), lichenization does not occur. In accordance with autonomous organisms, fungus and alga are sexually reproduced individually, even though they have also developed a system of combined asexual reproduction that consists of removing fragments of the lichen containing both.

The phenomenon of lichenization is supremely beneficial to the other organisms with which they share ecosystems: some types of lichenization (in which cyanobacteria take part as photobionts) are the largest producers of nitrogen salts (useful to plants) found in many ecosystems.

Fungi have also learned to establish mutually beneficial societies with plants, through the formation of mycorrhizas. In this case, the fungus *infects* its hosts' roots and extracts from their cells the carbon hydrates that the plant produces in its stem and leaves through photosynthesis. As a compensating factor, the fungus *lives* in order to improve the conditions of the plant's life. It increases the roots' capacity to absorb water and also facilitates the incorporation of some indispensable nutrients (mineral salts) for the plant, such as sodium, potassium, nitrogen, and especially phosphorous. Moreover, the fungus produces growth hormones for its host and protects it in the event of an attack of pathogenic microorganisms. This type of association is extremely widespread in the world of plants and allows many of them to survive in places where they would be unable to without help from the fungus.

Extraordinary cases of cooperation between insects and fungi can also be found. The wood wasp (*Sirex noctilio*) and the fungus *Amylosterum areolatum* maintain a relationship from which both mutually benefit. Female wood wasps have a few special receptacles, in the form of a pocket, in which they transport fungal spores. When they lay their eggs, they drill a hole in a tree trunk in order to construct small tunnels. They deposit their eggs in one of them and allow the fungal spores to fall into the others. It is obvious that the wood wasp greatly aids in the dispersion of fungal spores. The compensating factor takes place when the insect's larvae emerge from the interior of the tiny eggs. At that point, they find themselves with a good provision of food in the form of wood, decomposed by the action of the fungal hyphae that have proliferated before the birth of the young wasps.

As you can see, fungi display tremendous *sagacity* when it comes to exploiting the resources of their surroundings. The saprophytistic mode of life allows them to access a resource for which other organisms do not compete, like plants and animal life. Their greatest competitors in this field are bacteria, which, as we will see quite soon, have learned to keep under control. Many of those fungi that earn their way every day through parasitism have developed strategies based on cooperation with their victims, which allow the fungus and its hosts to be able to live in places that would otherwise be off-limits to them. In their relationships with other living organisms, fungi have demonstrated that they can be either extremely terrible enemies or supremely valuable allies.

There Is No Friend Too Small

As we have already commented, many bacteria are also saprophytes, whereby they compete with fungi for the source of food. This competition between fungi and bacteria has been going on for hundreds of millions of years, during which fungi have discovered a series of molecules that, having been released into the environment, eliminate their bacterial rivals and allow the fungus to dine alone at the banquet table. It is one of the oldest chemical wars of those that are known.

We humans are also at war with bacteria. They are the cause of many contagious diseases, the names alone make us shudder, such as leprosy, tuberculosis, syphilis, or bubonic plague. Moreover, bacteria are also responsible for infections that set in following wounds (in war, the workplace, surgery, or childbirth, for example), which have been the cause of death of large numbers of people. For this reason, many scientists have dedicated their lives to the search for substances that can eradicate bacteria without affecting our own cells. The task was quite complicated because it is not easy to find a poison that attacks one type of living organism (bacteria) without also acting upon another (our cells). Because of this, the cures that were developed were often almost as dangerous as the diseases themselves.

Fortunately, we have found an extremely valuable ally in fungi. Among the many researchers, who, in 1921, devoted themselves to studying the biology of bacteria, trying to find some efficacious method of fighting them, was a Scottish scientist named Sir Alexander Fleming (1881–1955). His method of work basically consisted of cultivating bacteria in circular, flat-bottomed receptacles called petri dishes. With the right culture medium, the bacteria proliferate in the petri dishes until they form colored colonies, perceptible to the naked eye. Adding the substances with which one wishes to experiment to the culture mediums in the dishes, it is easy to recognize whether the growth of the bacterial colony appears altered by the presence of the substance in question. Once a toxic product has been found for the bacteria, the next step is to experiment with its effect on mammals and, finally, humans. Obviously, it is much simpler said than done.

Working in this way, Fleming searched for substances that could destroy the bacterial colonies and, at the same time, would be innocuous to humans. Then he got an outlandish idea whose results had, in the long run, an enormous transcendence in the history of humanity: taking advantage of a cold he had, he added some of his own nasal mucus to one of the cultures. Once in contact with it, the bacterial colony disappeared. Fleming correctly deduced that nasal mucus must contain some substance that eliminated bacteria. He was also able to prove that saliva and tears produced a similar effect. Without a doubt, the substance present in mucus also existed in the other bodily fluids and must form part of a natural system of defense against bacteria. Fleming named this substance *lysozyme*. Unfortunately, lysozyme was only effective against a few bacteria, none of which were responsible for the terrible diseases and infections caused by them.

Seven years later, one September afternoon in 1928, Fleming again ran across a substance that was lethal for bacteria, but this time its efficacy proved much greater than in the case of lysozyme. In connection with an investigation linked to the flu virus, Fleming had sown several petri dishes with *Staphylococcus aureus* bacteria, a habitual tenant on the surface of the skin that is the cause of infections on the fingers when there are small wounds.

The technique for inseminating the culture mediums in the petri dishes is simple, but requires the necessary care to prevent the mediums from becoming contaminated by the spores of other microorganisms, ever present in airborne dust.

However, one of the preparations in Fleming's laboratory had not been carried out with sufficient care and was contaminated by the presence of a very common green mold (a type of fungus): *Penicillium notatum*. The fungi that belong to the *Penicillium* genus are found among the most common in everyday life. Some of them are the ones that cause spoilage to some fruits (especially citrus) in our refrigerators and pantries. They are also the ones responsible for the peculiar flavor of Roquefort (*Penicillium roqueforti*) and Camembert cheeses (*Penicillium camemberti*).

On that September afternoon, Fleming pondered the incorrectly prepared petri dish. His gaze, trained during many years in the observation of the most minute alteration in the growth of bacterial colonies, remained fixed on the area of contact between the mold colony and bacterial colony. Around the first, there was a space in which the bacteria did not flourish. Fleming immediately intuited what it meant: the fungus must have produced some substance that eliminated the bacteria.

Setting aside the rest of his research, Fleming concentrated on trying to isolate the substance produced by the fungus. He proceeded to remove the colony of mold from the petri dish and placed it inside a container filled with a culture medium. Soon thereafter, the fungus proliferated, and the culture medium turned yellow in color as a result of the presence of a substance synthesized by the *Penicillium*, which Fleming baptized with the name *penicillin*. His subsequent experiments showed him that penicillin was lethal for a great number of bacteria, even when diluted many hundreds of times.

Just five months after his first observation, Fleming made his findings known to the London School of Medical Research, but the news did not earn him the attention of his colleagues, and so Fleming proceeded to preserve his penicillin extract and went back to his other research projects.

Viewed in retrospect, we can feel astonished, and even indignant, at the lack of perspicacity of the scientific community of that period. But the reality is that it is one thing to discover a substance with therapeutic qualities and a very different thing to be able to produce it in sufficient quantities for it to be of use. At the time of Fleming's discovery, biochemical techniques had not reached the stage of development that they enjoy today, and the task of extracting, purifying, and isolating great quantities of penicillin remained outside their reach. That explains the apparent disinterest in which, in the end, the greatest discovery in the history of medicine was made.

Eleven years were to pass before science would again contemplate the existence of penicillin. In 1933, a brilliant researcher in biochemistry, Ernst Boris Chain (1906–1979), found himself forced to abandon his homeland, Germany, on account of his Jewish ancestry, and he decided to emigrate to Great Britain. After spending two years at Cambridge University, Chain set out for Oxford, where he began working with another notable scientist: Sir Howard Walter Florey (1898–1968), originally from Australia. Following a different line of investigation, they set to work researching bacterial substances produced by other microorganisms. This was how they became interested in Fleming's research on lysozymes, and this interest led them to discover his work on penicillin.

Equipped with more advanced technical means and being experts in the field of biochemistry, it did not take them long to isolate and consolidate a small quantity of penicillin. Their concentrate was a thousand times more potent than the extract obtained by Fleming. Next they experimented on mice and discovered that penicillin was extraordinarily efficacious against many kinds of bacteria, and it had no toxic effects for mammals.

With their meager supply of penicillin, the acquisition of which was quite laborious (2,000 liters of culture were needed to procure the necessary dosage to treat a single case), they attended some patients suffering from bacterial infections and were able to prove that the effect of penicillin was the same on humans as on mice. But the technical problem of producing penicillin in large quantities still remained. Florey and Chain sought help and resources from British industry and government, but it was 1940, and the country found itself immersed in the World War II and lacking the funds to devote to this type of research.

Then Florey traveled to the United States, which still had not entered the war, and there he found the support he needed.

Moreover, when the country found itself involved in the conflict, the government declared penicillin to be a high-priority wartime product. The research devoted to finding a way to produce penicillin in massive quantities intensified, and, near the end of 1942, the principal technical problems were already solved. Among the discoveries that made a spectacular increase in the production of penicillin possible was the discovery of another type of fungus, *Penicillium chrysogenum*, whose productivity was two hundred times greater than that of *Penicillium notatum*. This and other advances paved the way for the increase in the production of penicillin to such an extent that by the end of World War II, the pharmaceutical industry was supplying a sufficient quantity for the treatment of 7 million patients a year.

The discovery of penicillin not only made an effective cure for diseases with high mortality rates (like syphilis, diphtheria, scarlet fever, and gas gangrene, to give a few examples) available, but also opened the way for the later discovery of new substances synthesized by fungi, or other organisms, equally effective in the fight against deadly bacteria. The era of antibiotics had begun.

Fleming, Chain, and Florey were universally honored and their scientific achievements put on record with the awarding of the Nobel Prize for medicine in 1945. Although all three were awarded countless distinctions and honors, it was Fleming who received the most recognition. Among the most prominent degrees, medals, and appointments afforded him were the title of *sir* (in 1944), the Great Cross of Alfonso X the Wise (in 1948), the *Honoris Causa* doctorate from more than thirty universities around the world, and the title of honorary chief *Doy-Gei-Tan* of the Kiowa tribe.

If you, like us, are intrigued by the meaning of this last title (*Doy-Gei-Tan*), you will not mind if we put the subject of fungi aside for the moment in order to devote a few lines to the American Indians.

Led by our curiosity to know the significance of those words in Kiowa, we turned to our friend (and professor of linguistics at Madrid's Universidad Complutense) Enrique Bernárdez, an aficionado of the Amerindian languages who was familiar with some of them, like Navajo. But the problem of translating *Doy-Gei-Tan* was more difficult than we supposed. Today, the name *Kiowa* is applied to two different tribes: the Kiowa, properly speaking, who mostly live in the state of Oklahoma, and a branch of the Apaches, also called kiowas, but whose language is totally different from that of *authentic* Kiowas.

Our friend began investigating under the hypothesis that it was the *real* kiowa who had awarded Fleming the honor and found himself dealing with a nearly extinct language for which there is a grammar book but no dictionary. He then decided to get in touch with the Kiowa and discovered that almost no one speaks the language today. At present, the Kiowa tribe consists of around seven thousand members, and of that number, only about a thousand, the vast majority of which are old people, have any basic knowledge of their language. Unfortunately, Kiowa is among the languages whose disappearance seems inevitable in the short term.

Far from becoming discouraged, our friend continued investigating and put himself in touch with the director of the Kiowa Tribal Museum. Ernest Yellowhair Toppah (who was unaware of the honor that the Kiowa had bestowed on Fleming). Yellowhair convened a meeting of the tribal elders in order to consult with them, and among themselves, they solved the puzzle: *Doy* means *medicine*, and *Gei-Tan* means *he found*. So, at last, we were finally able to know, rescued from oblivion, the meaning of the title with which the Kiowa had honored Fleming: *The one who found the medicine*.

From the history of penicillin's discovery, we can extract a supremely important lesson for our time: no species is superfluous on this planet. Biologists and conservationist societies warn us daily about the catastrophe that biodiversity reduction, as a consequence of human activity, means for our planet. There are species that are fortunate enough to enjoy general appreciation (orangutans, gorillas, lynxes, bears, elephants, or whales, to give a few examples), and there is no need to argue the importance of keeping them on this planet. But there are other organisms that do not have our sympathy, or those we barely get to know because we snuff them out almost as soon as we discover them. We even suspect, with great certainty, that there are species we will never know, because we are making them disappear before we are even made aware of them. Leaving aside the ethical arguments, not because they are any less important, for curbing this bloodletting in the biodiversity of our world, the case of penicillin offers us an extraordinary example of how any living organism can be of great help to us. Even though we may not know how at the time.

The totality of this planet's living organisms constitutes our most valuable asset, but it also belongs to those who are to come after us. Let us be fair to them and not squander it.

Chapter VII

The First of "Us"

Darwin's Spring

If no vestige of organisms from the past was left on the planet, if fossils did not exist, how would we imagine the history of life?

Undoubtedly, the study of present-day organisms would have led us to discover the phenomenon of evolution. Certainly in Darwin's time, many fewer fossils were known than now, and their interpretations led some paleontologists to deny the reality of evolution. The veracity of evolution can be demonstrated from our knowledge of hereditary molecular bases, or through biogeographical explanations (that is, the geographical distribution of organisms) and comparative anatomy, such as that conducted by Darwin. The mechanisms that give impulse to the evolutionary process, random change in genetic material and natural selection, can also be studied without fossils. Moreover, evolutionary relationships among today's living organisms can be established with great reliability from anatomical, embryological, and molecular analysis, without dependence on the fossil record.

What would be denied us, were we not to resort to the study of fossils, is the knowledge of past life. How else would we come to know that more than 65 million years ago, some magnificent animals existed, populating the continents, oceans, and skies, and which then disappeared forever from the face of the earth? Surely, we would learn that chimpanzees and humans are closely linked (more so than the gorilla and chimpanzee), and we share a common ancestor who lived between 5 to 7 million years ago, but would we have any possibility of knowing the process by which we became human? Would we perhaps be able to know if it happened before the increase in the brain's size or the acquisition of biped locomotion? The answer is no.

Fossils not only offer us profoundly valuable information about the organisms that lived in the past and are no longer among us, but they also allow us to learn *how* the history of life developed. Without fossils, we might imagine that the diverse types of organisms that currently populate the planet gradually appeared over time. Perhaps we would conceive of another no less tranquil history of the extinction of intermediary organisms among the present-day forms.

What we certainly would not be able to imagine is the turbulent history narrated by fossils. Tragic episodes of instantaneous and massive extinction, glorious chapters of the sudden proliferation of numerous forms of different life, and long periods of calm, barely altered by a slow and gradual formation and extinction of species. Fossils allow us to know that the history of life has been convulsive and is marked by extraordinary events. From among all of them, perhaps the most fascinating is the appearance of our own world.

We have already said that when Darwin published *The Origin of the Species*, the fossil record was deeply limited, and voices from the field of paleontology arose against what Darwin defended and in favor of the immutability of the species. One of the arguments of greatest weight postulated by Darwin's adversaries was the sudden appearance of the animal kingdom, some 545 million years ago. This event serves to divide the history of Earth into two large chapters: the *Proterozoic* (or period predating animal life) and the *Phanerozoic* (or period of *visible* animal life). The first period of the Phanerozoic chapter is called the *Cambrian Period*. During Darwin's time, animal fossils had not been found in rocks predating the Cambrian Period (that is to say, *Precambrian* rocks), while they were found in great abundance in Cambrian rocks. Moreover, it was already possible to recognize in Cambrian fossils the first representatives of the practical totality of life forms that compose the animal kingdom: twenty-nine of the thirty life forms that are known today, and which have a fossil record, are already present in the Cambrian strata.

The theory of natural selection, formulated by Darwin, postulates the *gradual* appearance of different forms of living organisms and was scarcely compatible with the sudden appearance of the animal kingdom, with all its diversity already established from the outset, the way the fossil record seems to indicate. There were those who reasoned that this explosive appearance witnessed the moment of creation of animal life, as narrated in the Bible: in a flash, and all at once.

In chapter ten of *The Origin of the Species*, Darwin took up this problem, recognizing that it dealt with a question that was hard to explain, and one "can be properly put forth as a valid argument against the ideas expressed here." Although Darwin recognized, "I cannot provide a satisfactory answer to the question of why we do not find rich fossil deposits belonging to those periods which, presumably, occurred before the Cambrian system," he was sure that animal life must have existed prior to the Cambrian period. In support of this idea, beginning with the fourth edition of *The Origin . . .*, Darwin used the existence of an enigmatic fossil, called *Eozoon canadiensis*, which had been discovered, in 1858, by Sir William Edmond Logan (1798–1875), in Precambrian rocks in the western part of Canada.

However, the nature of this alleged fossil was highly controversial. For some paleontologists, it was a matter of an authentic fossil, which gave evidence of animal life's presence in the *Precambrian* Period, while many others, the Spaniard Juan Vilanova y Piera (1821–1893) among them, were of the opinion that it was a matter of a false fossil of mineral origin. The question was resolved by Karl August Möbius (1825–1908), who demonstrated the inorganic origin of the alleged fossil.

But apart from whatever the nature of the *Eozoon*, Darwin's principal argument for trying to justify the absence of a broad fossil record prior to the Cambrian Period was the fossil record's imperfection. In that regard, Darwin pointed out the following: "We must not forget that strictly speaking we only know a small portion of the world." This was Darwin's way of expressing his hope that perhaps Precambrian animal fossils would be found when the vast unexplored regions of the planet were better known. Putting it another way, Darwin hoped that in the future, in some part of the world, the *source* from which the river of the animal kingdom had sprung up—already teeming with water during the Cambrian Period—would be found.

Still today, the origin of animal life gets lost in the murkiness of the times, and although it is reasonable to suppose it must have occurred well before it suddenly burst onto the fossil record, what is true is that we have hardly any data on animal fossils until almost 600 million years ago. Perhaps the reason for this is that primitive animal life was not very complex, perhaps composed of only a few cells, and did not synthesize hard elements susceptible to fossilization. However, even keeping those factors in mind, the history we find recorded in sedimentary rocks speaks to us of a world dominated by unicellular organisms.

Stromatolites became frequent from around 2.9 billion years ago, and we find them in very different geographical areas, revealing the success of the bacterial mantles that produced them. It is conceivable that in the core of those mantles, suitable conditions were provided for the appearance of the first protists, heterotrophs (which could feed off the bacteria itself, or their remains), as well as autotrophs. We are reasonably certain that the first fossil belonging to a protist dates from about 2.1 billion years ago and is attributed to a unicellular alga (*Grypania spiralis*). It is also believed that unicellular algae were the organisms that produced a few relatively abundant fossils 1.6 billion years ago, the *acritarchs*.

However, there are no fossils of that antiquity that can be attributed with any certainty to the groups of heterotroph protists, among which must be the direct ancestors of animal life. The first tangible fossils of heterotroph protists come from strata that date back some 850 million years, in a section of central Australia known as Bitter Springs. What is known for certain is that unicellular algae secrete a sufficiently thick and resistant cellular wall that favors their fossilization, while many heterotroph protists do not, so that their possibilities of fossilizing and being recognized by researchers are much less.

Be that as it may, what is indisputable, as if this were a Wagnerian opera, is that the appearance of the first animal life in the fossil record was preceded by a colossal event of planetary proportions. Between 590 and 610 million years ago, what was surely the greatest of all the glaciations that our planet has endured took place. The great *Varangian* glaciation (also known as the *Laplandian* or *Marinoan*), not only lasted 20 million years, but affected the totality of the emerged lands, something unique in the history of Earth. Coinciding with the end of this Cyclopean glaciation, the imprints of a good number of never before seen organisms appeared, recorded in the sediments; among them, the first animal life.

On the outskirts of the Australian city of Adelaide, there is a place called Ediacara Hills, where an old abandoned silver mine is located. One fine day in 1946, the eminent Australian geologist Reginald Claude Sprigg (1919–1994) was in that very spot, carrying out a geological study related to mining, when he decided to take a short break for lunch. Without giving it much thought, he began turning over slabs of rock with his geological hammer and noticed that many of them displayed tenuous impressions that he recognized as fossils. Even though Sprigg thought that the Ediacaran rocks were the beginning of the Cambrian Epoch, it was soon established that they were in fact older and belonged to the end of the Precambrian chapter, with an antiquity of some 590 million years (almost 50 million years more than the oldest Cambrian rocks). Darwin's hope that Precambrian animal fossils might be found in some hidden corner of the globe had been fulfilled. By a marvelous coincidence, the term Ediacara derives from a word in the Aboriginal language that means *source.*

The Ediacaran fossils were not the first ones known from the Precambrian epoch, since some occasional discoveries had already been made beginning somewhere around the middle of the nineteenth century. But these were scattered findings of a few, hard-to-interpret fossils, which prevented scientists from tackling the problem of the origin of animal life. Nevertheless, the Ediacaran fossils, besides being quite numerous, record with acceptable clarity some of the structures of the animal life that produced them. The fact is that we are dealing with organisms that lack any type of skeleton, and what is preserved is the vestige of their soft tissues, something exceptional in the fossil record.

The fossilized organisms of Ediacara lived on the floor of a moderate-depth ocean. Periodically, the sands on the bottom were removed by storms and, on being newly deposited, covered the inhabitants that dwelled there. Thus buried, the organisms died and decomposed, but not before the shapes of theirs bodies left impressions in the sandy sediments that constituted their shroud.

The Ediacaran finding spurred many researchers to search for and meticulously record rocks of the same antiquity, paving the way for the discovery of about a dozen new deposits spread across several continents, confirming and broadening Sprigg's achievement in Ediacara. The group of fossils from all these deposits is normally referred to as *Ediacaran fauna*. The first studies of these organisms led to the conclusion that these are primitive forms of some of the phyla of animal life known since the Cambrian Epoch.

This point of view perfectly fit with what Darwin hoped: the apparently sudden appearance of animal life in the Cambrian Period had been preceded by intermediate stages in which the different phyla of animal life had been gradually surfacing. However, this interpretation of the Ediacaran fossils was contested by another much more revolutionary vision, proposed around the middle of the 1980s by the German paleontologist Adolf Seilacher, stating that many of the fossilized Ediacaran organisms do not correspond to authentic animal life, but rather to a new and different type of multicellular organisms which he baptized *vendozoa* or *vendobionts* (referring to the *Vendian* geological period, in which its fossils are found). According to Seilacher, vendobionts are differentiated from authentic animal life because they lack a mouth or any other orifice that allows them to ingest food, a characteristic that is distinctive of the animal kingdom.

As animal life, we differentiate ourselves from fungi, which are also multicellular, heterotroph creatures, in two fundamental aspects. On one hand, our cells are highly specialized in carrying out different tasks and are organized into tissues. Moreover, we take in food through external openings (basically, the mouth), and our cells then absorb the nutrients in the interior of an internal cavity (digestive apparatus), where the food is normally degraded by the action of substances released by specialized cells (except in the case of sponges and some types of internal parasites, whose morphology has varied a great deal as the result of their profound adaptation to that kind of life). This latter process is known as *digestion* and is exclusive to animal life.

Not having a mouth or other orifices, it can be added that vendobionts were organisms that lacked the ability to travel autonomously (therefore, they remained buried, because they could not escape), and their bodies displayed a large surface. Many of them are so strange that there is one author who came to propose that they were

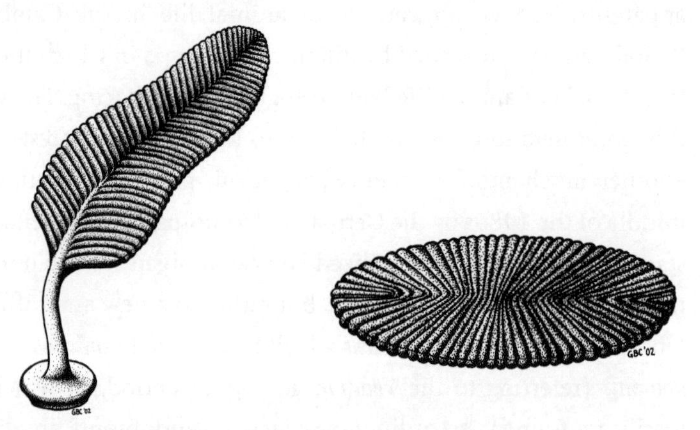

Figure 14: *Two Ediacaran vendozoa. Charniodiscus (left) and Dickinsonia (right).*

lichens. The absence of a mouth and inability to travel make it difficult to imagine how they were able to acquire food and forced

Seilacher to suggest that perhaps vendobionts directly absorbed nutrients from the sea, and they even included algae in their tissues and, like the fungi in lichens, obtained part of their food supply from the former. The imaginary schematic reconstruction of some of these strange creatures, such as the *Dickinsonia* and *Charniodiscus*, portray them as thin cushions, divided into compartments filled with liquid, where the algae that provided them their food probably lived.

This idea, that a type of heterotroph organism could have existed that acquired food by *cultivating* internal microorganisms and feeding them nutrients coaxed from the external environment, can seem outlandish and even ludicrous. However, in the current biosphere, there are animals that do exactly that. Among them is the tube worm, *Riftia pachyptila*, which lives at depths of over 3,000 meters in the immediate surroundings of underwater thermal springs, called funnels or fumaroles. These abyssal chimneys produce a stream of hot water, rich in hydrogen sulfide. We have already seen that hydrogen sulfide can be used as a primary source of energy by a group of chemolithotrophic bacteria for synthesizing organic matter.

Just like vendozoa, *Riftia* has no mouth or digestive tract; the only thing the animal has is a crest of gills through which it breathes, and this is located at one end of its body jutting out from the tube that houses it. Since it lacks a mouth, or any other opening in its body through which it can ingest the smallest particle, how does the *Riftia* get its food? The answer lies in a peculiar organ that occupies a good portion of the animal's body: the *trophosome*, whose cells are literally invaded by an extremely high number of bacteria capable of producing organic matter from the oxidation of hydrogen sulfide. Herein lies the secret of *Riftia's* survival: this *ingenious* worm takes in the hydrogen sulfide from the environment through its gills and brings it, by way of its circulatory system, to the bacteria that are swarming in the cells of its trophosome. In exact correlation, the bacteria pass along part of the organic molecules that they synthesize to their host. In this way, the cooperation between the worm and bacteria allows both to survive. This form of life has required highly complex adaptations in *Riftia*, which include the presence of a very special type of hemoglobin. *Riftia* is not the only animal that uses sulfuric bacteria for its survival. Other inhabitants of the abyssal fumaroles, like the clam *Calyptogena magnifica* and the mussel *Bathymodiolus thermophylus*, also do. So then, the modus of life designed for vendobionts is not as eccentric as it would appear.

Today, many paleontologists accept Seilacher's interpretation for a large portion of the Ediacaran fossils. However, there are other fossils that indeed can be related to some of the known phyla of the animal kingdom, like *porifera* (or sponges), *cnidarians* (sea anemones, corals, and medusae or jellyfish), *echinoderms* (starfish and sea urchins, among others), *arthropods* (the most numerous group of the animal kingdom, which includes, for example, insects, crabs, spiders, centipedes, and the exclusively fossil group of trilobites), and several other types, vermiform in appearance, which we commonly group under the heading *worms*.

This point of view has been bolstered in the last decade of the twentieth century following the discovery, in northern Russia, of the greatest deposit of Ediacaran fauna in the world. After the discoveries that were made in recent years in this region, it has recently been published that one of the enigmatic Ediacaran fossils, called *Kimberella*, could be the direct ancestor of *mollusks* (a phylum that includes, among other animal life, octopuses and related species, snails, and bivalves, such as the mussel or oyster).

In other words, around 600 million years ago, the direct

ancestors of some of the principal classes of animal life (of numerous species) lived alongside some strange creatures, vendozoa, which are now extinct. This conclusion opens the door to a question of tremendous interest: if one can trace the origin of some of the principal phyla of animal life as far back as 600 million years ago, when did their separation take place? That is to say: *when* did the first animal life appear?

We have already seen that the fossil record is not of much help for the period that extends beyond the Ediacaran epoch. Where do we look then? The answer lies in attempting to read the history that present-day animal life carries, written in its own genetic code. When a specific species produces, through evolution, two descendant species, both share (that is to say, all things being equal) the vast majority of their genes. The difference that separates them is still infinitesimal and only affects a minute fraction of their genetic material (the human and chimpanzee species diverged about 6 million years ago, and we still share 99 percent of our genetic material). From that moment on, the genetic divergence, like the morphological difference, continued to increase over time.

This fact allowed molecular biologists to devise an ingenious method, known as a *molecular clock,* for estimating the time that elapsed from the moment of separation of any group of present-day organisms. The molecular clock has been applied to many cases, a few times with greater skill than others. The use of the technique in question presents two difficulties, of which the most immediate is that of establishing the rate of genetic change. It is one thing to assume that change is produced in a consistent fashion and another to believe that the rate of change varies throughout time. Even in the simplest case, that of steady change, it is not easy to quantify the rate of change. If high values are chosen, meaning the change is rapid, the molecular clock will produce younger ages than if low values are used, meaning the change is slow. As one can see, it is possible to attain very different results depending on which rate is chosen.

Here the second difficulty arises. How are we to compare the results obtained using different rates of genetic change? Imagine that in using a specific rate of change, we reach the conclusion, for example, that two lineages diverged 10 million years ago, while if we use another more parsimonious rate of change, we conclude that the same strains separated 15 million years ago. How are we to know which of the two results is more accurate? The only possible way to answer this question is by using the fossil record. If fossils of both evolutionary lineages exist from 13 million years ago, for example, it is clear that the first result is erroneous, and the second one must be upheld. In other words, molecular clocks need to be *calibrated* based on the fossil record.

In the case before us, the origin of animal life, different molecular clocks have also been tried and have produced disparate results. In 1996, researcher Gregory Wray and his team published a study resulting from the differences and similarities found in a group of seven genes, present in all animal life, which indicated that the metazoa's point of divergence (in other words, the origin of animal life) was produced between 1 and 2 billion years ago, long before the Ediacaran Period. However, this result was questioned barely two years later by another study, headed by Francisco Ayala, who, after analyzing the disparities presented by different animals in eighteen genes, obtained a much more recent date, about 670 million years ago, for the separation of the two large groups of the animal kingdom: the *protostomia* (which includes, among others, strains of mollusks and arthropods) and *deuterostomia* (in which we are linked to echinoderms and vertebrates). Although the dates did not exactly correlate to the same event (since divergence among protostomia and deuterostomia was subsequent to the origin of animal life), and thus the results could be compatible, the divergence between both studies was too great. To further complicate the problem, a year later, in 1999, Blair Hodges and his group published the results from an even more extensive study, which substantiated the conclusion of Wray and his team, since they placed the line of separation between vertebrates (deuterostomia) and insects (protostomia) at almost 1 billion years ago.

The evidence contributed by the fossil record does not allow us to choose, strictly speaking, between the results offered by the different molecular clocks. The Ediacaran fauna dates back 590 million years and is subsequent to the points of divergence proposed by the different researchers.

There are those who believe that the date offered by Ayala and his team is too close to that of the Ediacaran fossils, and it does not allow sufficient time for the diversification observed in the fauna in question to have been produced; but there are also those who think that the date of 670 million years for the separation between the large groups of animal life is more compatible with the evidence of the Ediacaran fossils than the older dates.

Within this context, a finding published in 1998 by a group of paleontologists headed by Seilacher gains special relevance. These scientists claim to have found fossil evidence that established the existence of a *wormlike* animal in rocks from central India dating back approximately 1.1 billion years. This discovery consists of a configuration that they interpret to be the fossilized tunnel carved out by a worm that was excavating for food in the sediment in that remote epoch. It almost goes without saying that this interpretation has been debated time and again by other scientists. Nevertheless, should the existence of this type of animal life more than a billion years ago be confirmed, with more and better fossils, the die would be cast in this polemic, and the origin of animal life would date back in time far beyond what Darwin ever would have dared to dream.

Sponges, Jellyfish, and Worms

It may seem somewhat disappointing to you, having a *worm that fed on sediment* in your genealogical tree. But worms are a form of animal life that deserve our greatest respect, for they were the ones that *invented* many of the characteristics that we regard as being typical of animals. In order to place worms in the elevated position they deserve, let's look at the way in which the simplest animals earned their keep (and still do) during the Ediacaran Period.

The animals with the simplest anatomy that we know of today are sponges. Imagine yourself as an engineer of living things (like in the movie *Blade Runner*), and you are given the assignment to design an animal, the simplest one possible, to populate the ocean floors of a planet with conditions similar to those on Earth. In order to complete your task, you decide to eliminate the greatest number of organs possible and limit the number of cellular strains to the essential minimum. To do that, you must first decide what is the simplest way to acquire food, since as a good animal, your creature must be a heterotroph. After gathering sufficient information, you reach the conclusion that the least complicated form for acquiring nutrients is that of filtering water in order to extract any microorganisms and/or organic residue that are in suspension (arising from the activity of other living organisms) from it. Filtering the water, sufficient within the limitations of the designated assignment, provides the advantage of not having to move in search of food, since all that is needed is to have the water circulate through a filter. So far, so good, but you now need to design the simplest pump and filter possible.

Before commencing to devise a system with those characteristics, you decide to investigate solutions discovered by nature for the same purpose. When all is said and done, life has been subjected to a refining process, dominated by natural selection, for many millions of years, and you can certainly find some inspiration for starting your design. Then to your astonishment, you discover sponges, which provide you, not with an inspiration, but with the solution to the problem.

Sponges are the simplest animals in existence. They lack specialized organs, are *sessile* (that is to say, they are stationary), and consist of a few basic types of cells. But let's not be deceived by their simplicity; sponges are a marvelous biological *device* that have continued to function efficiently on this planet, where competition is fierce, for many hundreds of millions of years. A proof of the tremendous *professionalism* of these animals is the fact that more than three thousand different species of sponges are known.

As you have probably already deduced, sponges earn their way by filtering water, the simplest possible work for an animal. In essence, a sponge is nothing more than a sac with a large opening (or a few), called an *osculum*, and a multitude of small orifices or pores (from which they derive the name Porifera), called *ostia*. The sac's walls basically consist of two types of cells: those on the internal wall are collar cells or *choanocytes*, and those on the external layer are *pinacocytes*.

The cells of the internal layer are special because they exhibit a lengthy extension, similar to a whip (and is thus referred to as a *flagellum*), on the part facing inward to the sac's internal cavity. At the very base of the flagellum, encircling it like a small necklace (thus called collar cells or choanocytes), is a structure that reminds you of a little basket on whose frame the water-borne particles that constitute the sponge's food supply are retained. To prevent

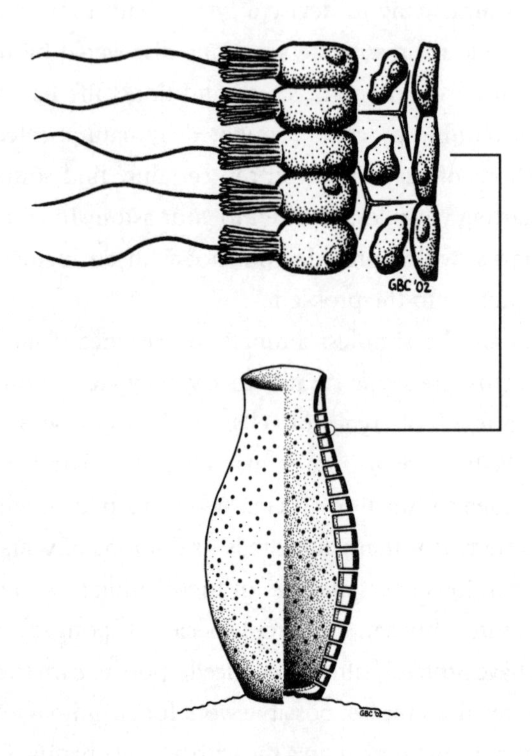

Figure 15: *Top: A diagram of a sponge showing the wall of the body, the canals situated among the ostia, and the animal's internal cavity. bottom: A detail of the wall of a sponge's body with the internal layer of choanocytes (on the left), external layer of pinacocytes (on the right), and mesohyl between both. Amebocytes and spicules are found on the mesohyl.*

already filtered water from becoming lodged in the vicinity of its filtering structure, the choanocytes beat the water with their flagella, constantly stirring it.

But besides these microcurrents, produced by the beating of the flagella, sponges need to generate a permanent stream of water that refreshes the water of its interior. The effort required for moving that mass of water is outside the reach of the choanocytes, and the sponge employs an *ingenious* trick to make the water cycle through. The sum of the diameters of all the ostia is much greater than the diameter of the osculum (or of the group of oscula). This funnel arrangement produces a *ventilating* effect that draws in seawater, making it circulate from the widest opening (the group of ostia) to the narrowest (the osculum). In this way, an infinite number of small currents of water continually pass through the sponge to converge toward the osculum, thus assuring a constant supply of fresh water for filtering. One of the simplest sponges, of the *Leuconia* genus, only 10 centimeters (4 inches) high and 1 centimeter (less than half an inch) in diameter, discharges and filters the respectable quantity of 22.5 liters (6 gallons) of water everyday (that is to say, it has a quantity of water almost equal to one hundred times its own volume moving through its interior). Obviously, this is a very efficient *device*.

But you, as an engineer of living things, are not easily impressed. You know quite well that capturing food is not enough for an animal to function properly. The extreme simplicity of sponges presents many problems for the adequate achievement of other functions and numerous doubts arise regarding the viability of such a simple structure. How can the sponge prevent the collapse of its extremely thin canals, which extend from the ostia to the osculum? How does the food, captured by the choanocytes of the internal layer, reach the pinacocytes of the outer layer? How does oxygen reach the cells, and how do they eliminate their waste? How can a sessile animal, such a seemingly appetizing and easy prey, defend itself against potential predators? How can such a simple animal reproduce if it lacks reproductive organs? To be certain the design works, you must find an adequate response these questions.

The key to answering many of these questions is found in a gelatinous layer that is located between the layer of pinacocytes and choanocytes of a sponge's body. This intermediary layer is called the *mesohyl* and is occupied by a group of cells that fulfill highly diverse functions. The principal type of these intermediary cells is made up of a few cells that are capable of modifying the external form of their bodies and shifting about in the core of the mesohyl: *amebocytes*. Amebocytes (or cells derived from them) are extremely busy cells and are given to profoundly disparate functions. On one hand, they are responsible for synthesizing skeletal elements, which serve to maintain the animal's shape and prevent the walls of the canals from collapsing. These elements can be needlelike splinters (called *spicules*) or fibers of a flexible substance rich in collagen. Bath sponges belong to this latter group, and what we are really using is the animal's fibrous *skeleton*. In the skeleton of a natural bath sponge (which can be bought in drugstores and pharmacies), you can clearly observe the network of little canals, just like the ostia and oscula.

Amebocytes also have the task of distributing the nutrients captured and passed along to them by the choanocytes, cell by cell; the amebocytes then issue these food particles to the rest of the sponge's cells (like a sort of deliveryman for miniature pizzas). The inhalation and elimination of food particles is the task of each one of the cells, which take oxygen directly from the water and release their waste product into it. The fact that the sponge is furrowed with plentiful currents of water assures it of proper regeneration, with the arrival of new, oxygenated water and the removal of water charged with waste. On the other hand, sponges depend on their needlelike splinters, some of them large, in order to keep many predators under control. To this *bulwark of lances*, some sponges add the ability to synthesize urticarial or toxic substances.

Sponges have two kinds of reproduction. On one hand, an asexual type of mechanism, consisting of the production of small lumps (or *buds*) on their surfaces, which contain every type of cell and the structure of a sponge. These buds grow and can break free from or remain joined to their progenitors, eventually forming, through this procedure, aggregates of individuals linked to each other.

Moreover, sponges are also capable of sexually reproducing. These animals are hermaphroditic but are never self-fertilizing because the sperm from one sponge is always going to find its way to other sponges. The reproductive cells (*gametes*) are formed, according to the species, from the cells of the mesohyl (generally, amebocytes) or from the choanocytes. The spermatozoon is released into the canals that run throughout the body of the sponge and is thus carried outside, while the ova remain inside the sponge. The sperm from one sponge enters the body of another via the current of water that penetrates the ostia. Fertilization takes place inside the sponge, and the embryo is retained during the first phases of its development inside the body of the *mother* sponge (which, let's remember, also acts as a *father* sponge, since its sperm will fertilize other sponges). During this period, the adult sponge not only shelters the embryo, but nourishes it through a special membrane.

Maybe this somewhat reminds you of how the maternal womb holds an embryo and actively feeds it through a structure solely engendered for that purpose. We placental mammals (and many other animals) do it too, even though in a completely different way (and which has nothing to do with the way in which sponges carry it out). As you can see, gestation (in its different forms) is a very old *invention* in the animal kingdom.

Thus you have a picture of what sponges are like, a solution as simple as it is brilliant to the problem of designing the simplest animal possible.

• • •

But let's try to take this line of harmonizing structural simplicity and functional complexity a bit further. In the end, sponges are limited to filtering water, but if we want to devise an animal that acquires its nutrients by devouring other living organisms, as in fact the vast majority of us do, maybe we will be obliged to complicating their structure a great deal more. The *role* of predator (which in a broad sense includes animals and vegetables alike as prey) presents new challenges, for which the structural simplicity of sponges cannot provide an answer.

First of all, it has to be capable of capturing organisms that will become the next meal. At this point, a dilemma is opened up: to take other animals as prey or to focus on plant life instead. The first possibility runs into the problem of the intended victim's ability to freely move (with the exception, of course, of sponges, which in this regard behave like plants) and, therefore, to escape. This difficulty does not exist where plants (and sponges) are concerned, and so the problem here is precisely the opposite: since the prey is not mobile, it is the predator that has to be able to move autonomously to find and gather food. In other words, food does not put itself at the table. Which of the two modi operandi, carnivorous or vegetarian, provides a simpler solution? Another group of fascinating animals offers us the solution to this riddle: the *cnidarians*.

On previous pages, we had occasion to make a brief presentation of cnidarians, among which are jellyfish, sea anemones, and corals. In fact, sea anemones and corals are a very similar type of animal generically known as a *polyp*, and so we can say that cnidarians include polyps and jellyfish. The latter are animals that we all know, and whose general shape could be described as that of *a sac with tentacles*. On the other hand, the morphology of polyps is less known. However, a polyp also is *a sac with tentacles*, even though in this case, the sac is affixed to the substrate, and the tentacles point upward. Basically, a jellyfish is nothing more than a polyp that is *turned upside down* and capable of floating.

As we have already said, a polyp is, in essence, a little sac with a single opening located on top. This aperture is called the *peristome*, and it works like a mouth. The internal cavity of the sac is called the *celenteron* and serves as a digestive apparatus. The walls of the sac are composed of only three layers: the external region (also called the *ectoderm*) and another on the internal part (or the *endoderm*), which is in contact with the celenteron. Also, as with sponges, both layers (the ectoderm and endoderm) are separated by an intermediary gelatinous layer that, in the case of cnidarians, is called the *mesoglea*. Surrounding the mouth are the tentacles, which are nothing more than thin extensions of the body's *sac* or central region, and they have the same structure as the latter.

As we have already said, the anatomy of a medusa is basically that of a polyp in an inverted position, so that its upper body, where the mouth is located, is pointed downward, and the lower part of its body, by which the polyp is affixed to the substrate, is now facing up. On the superior region of the body, the layer of mesoglea is greatly expanded, producing the medusa's characteristic bell shape. Since the mesoglea is less dense than water, this superior region acts as a float.

Up to this point, not many differences between the basic structure of a sponge and a cnidarian have been noted. The only thing that stands out is that the wall of the sponge's body is perforated by numerous little canals, something that does not occur in cnidarians, and the cnidarian's mouth is surrounded by tentacles, which are not present in sponges. The difference does not seem great. In fact, what really sets them apart and enables cnidarians to be predators is a special type of cell called the *cnidoblast*.

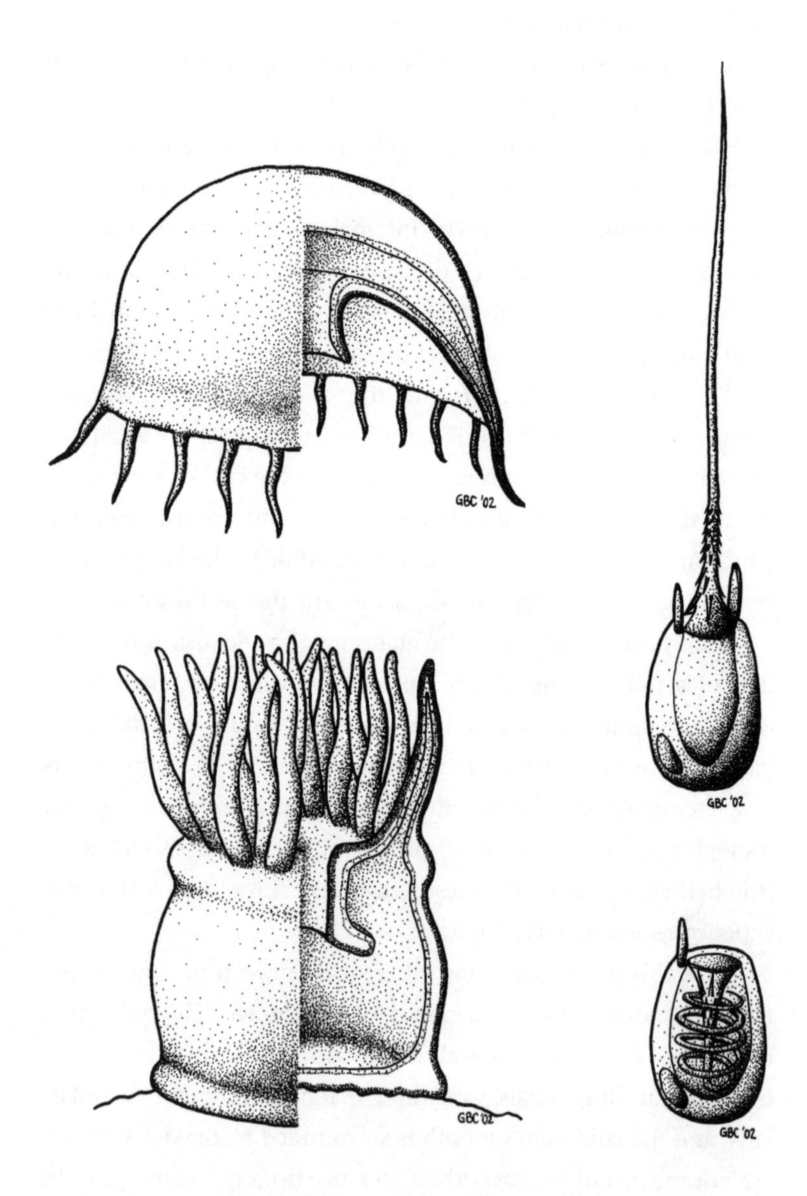

Figure 16: *Bottom right: a cnidoblast at rest, with its coiled internal filament and the cnidocilium on its apex. Top right: a discharged cnidoblast. Bottom left: a diagram of a polyp. Top left: a medusa dissected to show their internal anatomy.*

Cnidoblasts are located, above all, on the external layer of the cnidarians' bodies and are much more abundant on their tentacles. There are many different varieties of this type of cell, but they all share one fundamental trait: they are capable of discharging a filament that, in most of them, injects a toxic substance into any animal that brushes up against them. A vesicle, containing this filament, which looks like a coiled spring, is located on the portion of the cnidoblast that comes into contact with the external medium. This vesicle is connected to a thin extension of the cell (or the *cnidocilium*) that acts as a trigger. If something brushes against the cnidocilium, the filament is discharged and perforates the wall of the victim's body, injecting it with the toxic substance. In this way, any animal, large or small, that touches the tentacles of a cnidarian will provoke the discharge of a large battery of those miniscule cellular crossbows. The cnidarian's poison is potent enough to paralyze or kill its prey, or to produce lesions and even death in larger animals, such as unsuspecting swimmers.

But feeding off other animals takes more than killing them. Sponges capture small particles of organic residue from the water that their cells are capable of digesting in order to transform them into basic nutrients (glucids, lipids, proteins, etc.). But an animal, no matter how small it is, cannot be swallowed by a cell; first it must be digested. This digestion cannot be realized in the external medium, since the gentlest water current will carry away with it what has taken the animal so much work to capture. It becomes necessary to swallow the prey in order to be able to digest it. This is precisely the celenteron's function: to serve as a receptor where digestion takes place. Once the prey has been transformed into simple nutrients, the cells of the endoderm can absorb and distribute them to the rest of the animal's cells. Those remains of the victim that cannot be digested will be expelled into the exterior medium through the mouth itself.

But, as in the case of sponges, it is not enough to understand how an animal gets its food in order to assure ourselves of its viability. Again we can formulate numerous questions concerning respiration, elimination of waste, and reproduction. To these questions we can add some new ones: How do cnidarians swallow their prey? Are there differences among polyps, sedentary life, and free-swimming medusae in the way they function?

Let's leave the last question until the end and take a look at the other questions. As happens with sponges, respiration and excretion are independently carried out by each cell, taking in oxygen and releasing waste products directly into the water surrounding the animal, or rather the water that fills up the celenteron.

Cnidarian reproduction is one of the most surprising facets of the animal kingdom. Even though deeply marked differences exist in reproduction among the different groups of cnidarians, they can all be considered variants of a general pattern that is still found in many species. The principal and most extraordinary characteristic of this form of reproduction lies in a phenomenon known as *generational alternation,* in which one generation of polyps produces an offspring of different creatures, medusae, which, in turn, engender a new *crop* of polyps. To make this example of reproduction more peculiar yet, it suffices to add that polyps generate medusae in an asexual way, while medusae, which are male or female, use sexual reproduction for generating new polyps. Can you imagine a type of reproduction in which your children would be a different species of organism than you, and in turn, their offspring would produce a generation of grandchildren who would be once more like you? Add to that the idea that while you had to find a mate to carry out the perpetuation of the species, your children are producing offspring without a partner, from lumps on their bodies that grow in order to produce other living beings like you.

This astonishing mechanism of reproduction has been greatly simplified by cnidarians. Thus, numerous species exist that have reduced (or eliminated) some of the stages of reproduction. In these species, polyps only produce polyps, and medusae generate a few special polyps, reduced and with extremely short lives, whose only function is to generate new medusae. The large jellyfish that we find so bothersome at the beach, the sea anemones that we discover among rocks, or the polyps that form coral reefs, belong to these species with a simplified reproduction system.

In short, the most notable characteristic of cnidarians is not their peculiar mode of reproduction, but rather a capacity shared by all animals and nonexistent in sponges: cnidarians move around. They exercise this ability in both senses of the word recognized in the dictionary: they are able to move parts of their bodies and also have mobility. It would be meaningless for a cnidarian to be able to capture prey, through the discharge of its cnidoblasts, if it were not capable of swallowing it in order to digest it and absorb the nutrients. This very simple phenomenon, the swallowing of food, requires two things: the animal make use of muscular cells arranged in fibers[4] and these fibers work in coordination, thanks to the action of nervous cells, two of the most characteristic components of animals and something we find for the first time in cnidarians. Their hairlike tentacles contain longitudinal muscular fibers (in other words, arranged along the length of the tentacle), whose contractions impel its graceful movement. Surrounding the mouth are other fibers that allow the animal to close and open it at its convenience. There are also fibers on its

4. The muscular fibers of cnidarians differ from the musculature of all other animals in their structure as well as in their origin in embryonic development.

body, arranged longitudinally as well as circularly (that is to say, around its body), which allow it to produce movements that contribute to its ingestion of food, or which enable the animal to move away in the face of danger. Diverse combinations in the action of the different fibers allow individual polyps to realize a broad spectrum of contorted movements, thus giving the animal mobility along the ocean floor.

The coordination of all these movements is the responsibility of a diffuse network of nervous cells that are dispersed throughout the body and tentacles.

The existence of muscular fibers and nervous cells has enabled specific types of cnidarians to abandon the ocean depths, from which sponges cannot escape, and freely travel. Likewise, medusae benefit from their ability to change the shape of their bodies in order to generate a water current that they use to propel themselves. But the freedom of movement that medusae enjoy presents them with a series of problems that are unimaginable for a polyp that lives attached to the substrate. Medusae live in a three-dimensional world, while the world of polyps has only two dimensions. When medusae swim, they experience the action of a physical phenomenon unknown to polyps, which hardly move at all: *acceleration*. Moreover, a medusa needs to answer a question that makes no sense to a polyp: "How am I positioned, face up or face down?" (medusae have no head, but let's allow them this license). Being mobile is so complicated!

How have medusae solved all these problems? The answer is simple: thanks to the presence of sensory organs. These organs are exclusive to animals, and they provide us excellent information on the environment and our relationship with it. The characteristic sensorial organ of medusae is called the *rhopalium* and, in fact, consists of a group of several sensorial structures. On the other hand, there is a small fossette whose wall is covered with special cells capable of analyzing the chemical composition of seawater; it is an authentic olfactory organ.

Located in another region of the rhopalium is an extremely peculiar organ called the *statocyst,* which consists of a miniscule granule of calcium carbonate surrounded by sensory cells. When the medusa moves or is pushed along by a wave in a particular direction, the little granule of the rhopalium, which is not firmly connected to the body of the medusa, moves more slowly as an effect of its inertia. This distinct movement of the tiny granule brings it into contact with a specific area of the layer of sensory cells, which detects the intensity (acceleration) and direction of the movement. Thanks to this information, the medusa *knows* its spatial orientation and if it is *face up* or *face down,* enabling it to correct its position at all times.

The medusa's statocyst is an authentic *system of inertial navigation,* very similar to the one that allows pilots to know, at any moment, the position and speed of their planes. Here is an example of an inert system: imagine yourself standing in the center of a Pullman car that has no windows. You'll immediately know when the train starts, because as an effect of its inertial state, you'll be thrown back toward the rear of the car. You will also know whether the train has started off at a slow or fast speed depending on the force with which you're hurled backward. If the train stops, you will be impelled forward, again with a force proportional to the intensity of the braking. If the train rounds a curve, you will find yourself leaning toward the opposite side of the aisle in accordance with the sense of orientation. However, in this example, you are playing the same role as the medusa's little granule of statocyst, and the walls of the car, which you repeatedly bump against, are equivalent to the layer of sensorial cells.

But not only medusae and modern planes are equipped with an inertial system: on the interior part of our ears, we make use of another very similar mechanism, which also allows us to know whether we are moving, in what direction, and with how much force.

In addition to these sensorial organs, many medusae are also able to detect light through a few sensory cells (*photoreceptors*) that are grouped together, in the core of the rhopalium, in an organ called the *ocellus*. Since light in the sea always comes from above, ocelli are also quite useful for letting a medusa know in which direction it is moving or how it is positioned.

But the really astonishing part, with regard to sensory organs, comes by way of a special type of medusa called the *cubomedusa* (also known as the box jellyfish), owing to its peculiar shape. This animal has real eyes, equipped with a cornea, lens, and retina that *would allow* it to see in a way very similar to how we see; that is to say, forming images of the external world. We say *would allow* it because the cubomedusa, like all other cnidarians, has no central nervous system where information captured by the eye can be processed. Why then does the cubomedusa have eyes, when there is no brain that can *see* the captured images? It is a highly intriguing question, but up to now, there is still no answer for it.

As you can see, cnidarians are remarkable animals, capable of providing solutions, from their structural simplicity, to extremely complex problems. Before leaving them, to take a look at other no less fascinating animals, we want to again mention some of their achievements, which are really extraordinary. For example, the *Cyanea arctica* species is a terrifying medusa that can weigh up to 1 metric ton and whose tentacles can exceed 3 meters (almost 10 feet) in length, which makes it one of the largest invertebrates to ever inhabit the planet. Among cnidarians of the polyp class, these members of the *Branchiocerianthus* genus, which live at great depths of up to 3,000 meters (9,842.5 feet), can reach 3 meters (9.8 feet) in length. An appreciable size for a single polyp. Still, the most spectacular feat achieved by cnidarians is the teamwork of millions of individuals of species much more modest in size: these are the polyps that form reefs. Their most colossal work is the Great Barrier Reef, which can be seen from the moon, located along a good stretch of the eastern coast of Australia. There is only one other structure built by an animal from our world that is also visible from our satellite: the Great Wall of China.

Sponges and cnidarians are the simplest animals that populate our planet's oceans, which, straight away, also makes them ideal candidates for being the most primitive animals, since it is reasonable to say that the simplest precedes the most complex in the history of life. In fact, found among the organisms represented in the Ediacaran fossils are diverse forms of cnidarians. However, and despite their structural simplicity, sponges and cnidarians are too specialized, in their anatomy and form of life, to represent the type of animal that gave origin to the rest.

A Treasure in the Sediment

Most animals that form part of our daily lives, and also those that we know through books, documentaries, and museums, have a general body structure with which we are very familiar. Insects, worms, crabs, octopuses, kangaroos, fish, birds, frogs, or dinosaurs all have a *head* and display *bilateral symmetry*. The head is easy to distinguish because it is situated on the upper extremity of the body; it includes a mouth, most of the sensory organs (like the eyes), and houses a brain. By bilateral symmetry it is understood that organisms are constructed in such a way that half of the body is symmetrical with respect to the other half (in other words, like your image in a mirror). Of course, the correlation between both halves of the body is not always exact (we all have one leg a little longer than the other, for example), but it is such a salient feature that biologists divide animals into *bilateral* (with bilateral symmetry) and *nonbilateral.*

The nonbilateral group is heterogeneous because it includes animals without any sort of symmetry (sponges, for example) along with others of radial symmetry (cnidarians). To further complicate matters, among bilateral animals, there are cases like that of the starfish, which has a bilateral symmetry in its larval stage, but changes its form of symmetry in the adult state.

The human body has only one symmetrical plane (as happens with all animals of bilateral symmetry), while the circle has multiple planes, which crisscross in the center like the spokes of a wheel (on a bicycle, for example). This kind of symmetry, in which an animal's body can be divided into equal halves through several symmetrical lines (which crisscross in the center), is called *radial symmetry*.

If we slice a polyp on a vertical plane that runs from one of its tentacles to the one that is situated straight across on the opposite side of the body, we will end up with two similar halves (symmetrical), and since the animal has many tentacles, it also exhibits numerous symmetrical planes that crisscross at its center.

Symmetrical classification is a concept of great importance, because as we have already stated, animals are divided into two large groups: those that exhibit bilateral symmetry and those that do not. Why are some animals structured one way and others another way? To make the question more intriguing, it should be added that bilateral animals (almost always[5]) have a head, while

5. In biology, almost all norms have their exceptions. Thus, for example, bivalves, which are the group that includes mussels, scallops, cockles, oysters, and the like, constitute a category of animal that exhibits bilateral symmetry but without a head, even though their direct ancestors did, in fact, have one. The curtailment of the latter in bivalves is an adaptation to their modus of life.

nonbilaterals do not. Is there some relationship between the presence of a head and symmetrical classification? Since the head is the part of our body of which we humans are most proud, the answer to those questions holds a special interest.

The key to all of this resides in the modus operandi of one group as opposed to another. Nonbilateral animals (sponges and cnidarians) trap their food rather than move toward it. For a sponge or polyp, there is only one spatial orientation of importance: the *upside-down* axis. Below it lies the substratum, to which the nonbilateral must attach itself if it is not to be dragged away to a place where food cannot reach it, while above the surface, there is food as well as predators. Therefore, these animals have an inferior region different from their superior region, each one geared to a different task (this situation applies equally to medusae, except for an inversion of terms: for them, food lies below and the equivalent substratum above, since they are free-floating animals).

Apart from the difference between what is up and what is down, all other spatial orientation is similar to the point of view of sponges and cnidarians: food can reach them from any location. It is for that reason that sponges have pores all over their bodies and cnidarians have tentacles pointed in every direction: no region of space is better than another. If you were out fishing on the water and had several reels, surely you would place them all around the vessel, because fish can come from all directions.

But passively waiting for food to reach it is not the only way for an animal to secure its acquisition, just the simplest method of procurement. Obviously, another possibility also exists: the active search for the food. But in this case, radial symmetry is not the best solution, because now not all spatial directions are equivalent. Besides the up-down axis (which continues to be important in animals that move toward their food, because in the final analysis, the ground is not the same thing as the sky), another axis of prime importance for the life of the animal now comes into the picture: the *front-to-back* axis. The animal no longer waits for its food to arrive from any number of directions, but instead moves forward to where it will always find it lying ahead. This situation calls for the need to develop an anterior region specializing in two principal tasks: to locate and swallow food. To adequately fulfill both, it is necessary for the sensory organs to be concentrated in the anterior region (to guide the animal toward the food) and for a mouth to be located in that anterior region (to swallow the food).

Since, in order to reach food, it is necessary to analyze the complex information proceeding from the sensory organs and to react accordingly, it is highly practical to concentrate one part of the nervous tissue in this anterior region. Thus, the active search for food also determines the formation of a head (or process of *cephalization*): a specialized region of the body, located in its anterior, which includes the sensory organs, mouth, and a concentration of the nervous system: the *cephal*.

For its part, the posterior region of the body is the ideal location for the placement of the canal that exits the digestive tract, the anus, so that the food that has already been processed will not pass through the mouth again.

Thus, in the bodies of animals that search for their food, two specializations with regard to the up-down and front-to-back axes have been produced. Spatial regions that are neither up nor down, nor forward nor backward, make no difference to these animals: food can be either to the right or left, and therefore, no distinctions are established between the two sides of the animal's body, which can be similarly constructed. It has been the dynamic form of life, characteristic of most animals, that has determined the existence of a head and bilateral symmetry, those features that are so familiar to us.

If you stop for a moment to think about it, it will be obvious that animals are not the only objects in our daily lives with bilateral symmetry and a *head.* Almost all the machines we humans have designed with the specific goal of dynamic motion (automobiles, trucks, buses, trains, airplanes, ships . . .) are constructed along a front-to-rear axis, with an anterior region different from the posterior region, and their right and left sides of equal design. In the majority of cases, the section housing the *intelligence* apparatus (where the drive mechanism goes) is located in the anterior region. If our machines also have bilateral symmetry and almost always a *head,* it is not because the engineers who designed them meant to copy animals, but because it is the best design for actively moving from one place to another.

Still, today there is another characteristic that almost all animals have in common, but is more difficult to perceive because it has to do with our internal structure. It is the surprising fact that, apart from the digestive tract, which runs through us (in fact, from mouth to anus), our bodies are excavated with numerous ducts and cavities, to the point that, for the most part, we are hollow.

Yet, the inverse situation would be truly surprising, that is to say, if the bodies of animals were solid. In that case, the internal organs would be imprisoned and, as a consequence, subjected to strong compressions every time we bent over. Moreover, there would be no free space for placing in the organism's interior the networks of ducts that constitute the respiratory, vascular, excremental, and reproductive systems of the different animals.

To tell the truth, not all animals need hollow spaces in the interior of their bodies. In the case of sponges and cnidarians, their simple, two-layered structure allows them to live without any need for a respiratory, vascular, or excremental system. We have already seen how simply they carry out the functions that, in more complex animals, are performed by the organs and internal systems. On the other hand, neither sponges nor cnidarians stand out in the animal world because of their ability to double up their bodies, and the type of limited movement exhibited by polyps and medusae is conveniently absorbed by their intermediary gelatinous layer, the mesoglea. However, this absence of internal space is only possible in very simple animals. An increase in complexity requires the existence of hollows in the body's interior.

In most animals, the ducts and cavities are lined with a special tissue called *coelom* (or *coelomate* cavities and ducts). Coelom is not the only solution to the problem of acquiring empty space in the body's interior. Some groups of animals (quite simple in their anatomy) have found alternative solutions, and biologists call them (depending on the type of solution) *acoelomates* (that is to say, without coelom) and *pseudocoelomates* (or with false coelom). The explanation of the alternatives represented by the acoelomate and pseudocoelomate groups would take us far afield from our main topic and unnecessarily complicate this chapter, and besides, most animals are coelomates.

Since the majority of animals, including us, are coelomates, having a head and bilateral symmetry, we can reasonably assume that we have all inherited these characteristics from some specific type of animal, our most ancient common ancestor. Given that, as we have already said, neither sponges nor cnidarians are good candidates for occupying such a position, where do we look? This is where *worms* come into the picture.

Under the heading of *worms*, we are familiar with a very disparate group of invertebrates that only have their external aspect in common. Most of them (caterpillars and silk worms, for example) are, in reality, larval states of some types of insects, like flies, butterflies, or moths. Under that classification, we also include many other different types of animals, among which the best known are *annelids* (like earthworms and sea worms) and *platyhelminthes* (such as tapeworms and ribbon worms). As you see, not all worms are alike.

Platyhelminthes are a group of supremely interesting animals, because although having bilateral symmetry and, to a greater or lesser degree, a head, they lack coelom. This combination of traits is highly primitive and makes many scientists think that the first animal with bilateral symmetry must have been quite similar to some present-day platyhelminthes. Many of them (like the tapeworm) adapted themselves to living as parasites inside other animals, and their anatomy became extraordinarily simplified, and thus they are not good candidates for representing the ancestor of the other bilateral animals. But there is a group of free-living platyhelminthes that can help us to understand what the ancestral bilateral animal was like: these are known as *turbellaria*.

Turbellaria (flatworms) are aquatic animals, with a cylindrical or dorsoventrally flattened body (which is to say, from top to bottom) some 5 centimeters long, although some species grow no more than half a millimeter in length while others reach a length of 60 centimeters. Most are carnivorous and actively pursue their prey. To do that, they have a wide panoply of sensory organs, which includes *chemoreceptors* (in other words, the sense of smell), tactile cells (the sense of touch), and an ocellus (a simple eye). In the majority of them, the ocelli only distinguish shadows and light, but in a small group, these organs are authentic eyes that can come to form imprecise images. Flatworms move in a distinctive way; the cells of the external layer of their bodies, the epidermis, are equipped with small extensions, like little hairs, called *cilia*. (Do you remember the cnidarians' cnidoblast? That type of cell also had a cilium, called a cnidocilium.) The rhythmic vibration of the thousands of cilia of the epidermis propel the small flatworms, as if they were the oars of a galley.

In order to capture their prey, turbellaria make use of adhesive organs on their heads and posterior regions of their bodies. Once the prey has been located, turbellaria adhere themselves to it by the head and to the substratum using the posterior region. Then a terrible body-to-body struggle ensues in which the predator (according to the species) can swallow its prey whole, consume it little by little, or inject digestive substances into its interior. In some species of turbellaria, the females do not have a copulatory orifice, and so the penis of the males has a hardened structure, called the *stylus*, designed for perforating the wall of the female's body and injecting it with sperm. This stylus is even used for stabbing prey, once it becomes stuck to the surface of the predator.

In turbellaria, we can see the simplest model of a bilateral animal, perfectly equipped for leading a free life and finding food. But, as we have already said, this strain of animal lacks coelom, which leads us to ask, how did coelom manage to emerge in the course of evolution? Without a doubt, the presence of cavities and ducts is necessary for animals more complex than turbellaria to be able to exist, but what was the initial advantage in having it? What opened the way for the appearance of more complex animals, among the ones we know? Evolution isn't a process that acts over a long period of time, for the advantages of a structure have to be evident the very moment it makes its appearance; natural selection works in the present, not in the future. All in all, in the absence of coelom, turbellaria do just fine in the present world.

Different responses have been offered regarding the question of what was the advantage that coelom conferred on animals as simple as present-day turbellaria. There are those who believe that being equipped with cavities allowed for improved conveyance of oxygen and carbon dioxide between the body's interior cells and exterior medium. This hypothesis is highly reasonable, since turbellaria consume much more oxygen (up to ten times more) than simpler, less complex animals, like cnidarians, and the existence of a circulatory system would make the transportation of this gas much more efficient. But there is a still simpler hypothesis for explaining the appearance of coelom: the advantages of having a *skeleton*.

When we speak of a skeleton, we are accustomed to thinking of a group of sturdy components that shelter or support animals. However, the word skeleton carries a broader meaning in biology: it simply has to do with some structure that serves to sustain or confer consistency on the animal's body, whether composed of hard components or not.

Within this context, it makes sense to speak of a *hydrostatic skeleton;* that is to say, a skeleton comprised of a liquid. Even though, at first, it may be that liquids are not a good material for conferring solidity on the bodies of animals, what is true is that they are ideal for that very purpose. If a biological structure is filled with a liquid to sufficient pressure, it attains a remarkable consistency, and even hardness. Besides faithfully fulfilling this mission, the hydrostatic skeleton offers the advantage of being fluid; in other words, it can move from one side to another of the animal's body, allowing it to change form or consistency.

Sea worms of the *nereid* group are the ones we normally use as bait for deep-sea fishing. These are carnivorous animals that actively capture their prey. To accomplish this, they make use of their hydrostatic skeleton, which gives them mobility and enables them to trap their victims. They swim or crawl along the bottom thanks to the undulating movements (*serpentiform*) of their bodies, enacted by changes in the distribution of the liquid that fills the cavities of the body.

But the most dramatic use that nereids make of the great versatility of their hydrostatic skeleton lies in the way in which they capture and swallow their prey. The anterior section of their digestive tract, the *pharynx*, is normally invaginated toward the interior (like the finger of a glove if we push it inside the glove itself). At the tip of the pharynx is the mouth, which is armed with a pair of imposing, rigid jaws that, in a state of repose, are hidden in the interior of the animal. When the apparently harmless nereid wants to capture a prey, the musculature of the body's wall that surrounds the pharynx contracts and compresses the liquid, which acts like a hydrostatic skeleton. As a result, the hydrostatic skeleton functions like a pincer on the pharynx, causing it to violently thrust outwards toward the exterior medium (like the finger of the glove when we forcefully blow inside it) with its jaws at the end. Once the food has been captured, the pressure on the liquid decreases, and the pharynx is again retracted inward by the action of the corresponding muscles, pulling the prey inside.

As spectacular as the kinetic strategy of nereids may seem, we must not forget that turbellaria, which have no coelom (nor, therefore, a hydrostatic skeleton), also manage quite well to move and capture food, and thus the appearance of coelom must have served some other purpose than that for which turbellaria are perfectly able to do on their own. Here it is in a nutshell: what can annelids do, with their coelom (and let's remember, this is their hydrostatic skeleton) that is beyond the scope of turbellaria? The answer is dig.

All the different types of animals we have seen up to this point are aquatic and probe the water for food, whether it is floating in it or on the surface of the substratum. But another potential source of food exists in marine ecosystems; it is found beneath the substratum, in the sediment. The organic particles that are suspended in the water, as well as the organisms that live in it (swimming, floating, crawling, or affixed to the substrate), end up spread across the ocean floor, where they are gradually buried beneath a rain of sediment, which the rivers (and the wind, to a lesser degree) drag out to sea. In this way, a large quantity of organic material continues to accumulate beneath the surface of the ocean floors. This food is available to whoever can access it.

The most efficient way for an organism to extract organic material from the sediment consists of swallowing it and then separating the food from the part that is not digestible. The best way to do it is to move inside the sediment, swallowing it through the mouth and expelling the waste through the anus. Obviously, this requires the animal to have an elongated form (in other words, a front-to-back axis and, therefore, bilateral symmetry) and to be able to move actively.

Turbellaria fulfill these conditions, but their distinctive method of motion, through the action of cilia, is inoperative for getting around inside sediment, which is much denser than water. Moreover, and this is the most important point, in order to pierce the sediment, the body must be firm; in other words, it must have a skeleton. Annelids, with their hydrostatic skeleton, are perfectly equipped for boring holes into the substrate and searching for food there. The earthworm, for example, survives exactly in this way in the core of our fields and gardens. The appearance of coelom could very well be, originally, an adaptation for digging tunnels in marine sediment and, in this way, accessing a new kind of resource.

Thus it is that among modest *worms,* we can find good candidates for upholding the honor of having been the first to develop the basic traits that characterize the majority of animals: a head, bilateral symmetry, and coelom. Were it to be confirmed that the discovery announced by Seilacher and his collaborators, that they have found a fossil worm tunnel in rocks older than 1.1 billion years, we would be standing before the most ancient trace of the presence of an animal bearing those three attributes.

Some Hard Types

The reconstructions made of fauna from the Ediacaran Age often speak to us of an *idyllic* world, more tranquil and peaceful than that of later periods. The absence of any type of defensive element in the anatomy of the vendazoans, such as their limited movement ability, suggests to us the existence of a world free of nasty predators. On the other hand, we have already seen that neither does it appear that their way of life was based on devouring their neighbor. It is true that carnivorous animals, of the cnidarian and turbellaria type, must have existed, but their predatory abilities were limited to small prey, just like today. The Ediacaran world also included other equally peaceable beings that searched for their food by filtering water, taking in sediment from the ocean floor, or *grazing* on the abundant microbial mantles that formed stromatolites (supremely abundant during that period).

This *placid* world abruptly disappeared from the face of the earth about 550 million years ago with the extinction of the vendazoans. The microbial mantles even continued flourishing and producing stromatolites, along with cnidarians and excavating animals, for a few million years more, but their hours (or rather, their millions of years) were also numbered and soon declined in an abrupt manner until reduced to the present situation. Just when this happened, new types of animals appeared, which had access to a revolutionary novelty: they were equipped with *mineralized skeletons*.

Mineralized skeletons are formed by hard, rigid pieces composed of mineral salts like calcium carbonate and calcium phosphate. This type of skeleton is extremely efficient when it comes to giving consistency to the animal, although, unlike the hydrostatic skeleton, it does not allow it to modify its shape. Nevertheless, other advantages compensate for this limitation many times over. In the first place, mineralized skeletons allow external structures to be built that isolate the animal from the external medium and protect it from the attack of eventual predators. Moreover, the rigid pieces offer wonderful surfaces for muscular action, which allows the animals to carry out very diverse activities in a much more efficient way. Among them are locomotion and the ability to trap and process food.

Animals with rigid skeletons entered the scene in a very modest way about 545 million years ago (from this point on, and except for a clearly specific mention of them, we will identify *mineralized skeletons* simply as *skeletons*). The first evidence of this new type of animal consists of a few miniscule cones of a chalky nature, called *Cloudinia*. Barely 5 million years later, at the beginning of the Cambrian Epoch (in the period known as the *Tommotian Age*), animals with a skeleton had become somewhat more diversified. Among them were the *Aldanella*, a few tiny seashells twisted into a spiral like snails.

Although somewhat more varied, these skeletons continue to belong to animals greatly condensed in size. The great event gave no hint that it was about to burst upon the scene. In barely 10 million years (a bolt of lightning in the history of life), the seas were populated with new animals, larger than the ones that preceded them, and equipped with very different skeletons. Then came echinoderms, new forms of mollusks, *brachiopods*, and *archaeocyathids* (a group of animals similar to sponges and the first reef formers), among others. A little later (in the next chapter on the Cambrian Period, called the *Atdabanian*), the first arthropods appeared: *ostracods* and *trilobites*.

It is important to point out that surely the latter (ostracods and trilobites) were neither the first nor the only arthropods at the beginning of the Cambrian Period. All arthropods share the presence of a hardened skeletal covering, but in the majority of them, it is not mineralized, and therefore, their fossilization is highly improbable. The skeletal covering of arthropods is composed of a substance of organic nature, *chitin*, which, although it can be extremely hard (think of the shell of a beetle or cockroach), easily decomposes when not mineralized. Only some arthropods (like trilobites, ostracods, or present-day crabs) add mineral salts to their *chitinese* shell, thus facilitating their fossilization.

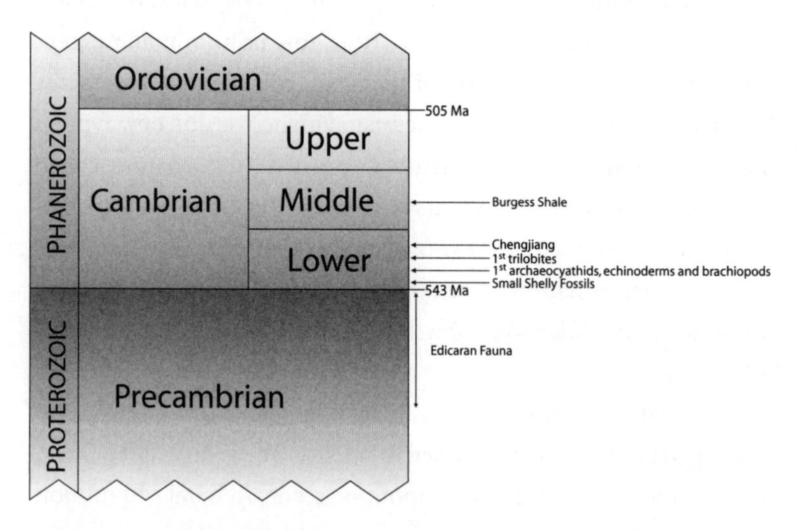

Figure 17: *Chronostratographic outline of time sequences of the main events in the history of the first animals. Also shown is the position of the Ediacarian sites, Chengjiang and Burgess Shale. The expression "Small Shelly Fossils" refers to the strata with the remains of small mineralized skeletons.*

The echinoderms of long ago were quite different from the ones that populate our oceans today. But just like today's echinoderms, their predecessors had a skeleton formed by plates of calcium carbonate. These plates are situated beneath the animal's epidermis and is therefore an internal skeleton.

Brachiopods have an external shell formed by two valves, just like *bivalves* (mollusks and company), and they are so similar to each other externally that to learn to distinguish between them is one of the primary tasks of students of paleontology. Without any doubt, brachiopods survived by filtering seawater, and most probably, that was also the modus operandi of most of the first echinoderms.

But the real main characters in the Cambrian seas were the trilobites. These animals are similar to the present-day king crab in their external anatomy. Its body is covered with an articulated skeleton (something like medieval armor) and composite eyes (like those of real insects), and what is most important, they had *legs* (or, if you prefer, articulated appendages). In the history of life, only two types of animals have appeared with legs: arthropods and terrestrial vertebrates. Legs are appendages that are sustained by the presence of a rigid skeleton (external or internal), divided into articulated segments among themselves.

We have already stated that the skeleton offers a magnificent surface of support for the muscles to carry out their activity, which facilitates the actualization of vigorous movements. The advantage of having the skeletal extremities articulated lies in the fact that it permits the relative mobility of its parts. If you ever had your leg in a cast, you definitely know the difference between an articulated skeleton (our bones) and a rigid one (a plaster cast). It is not an accident that the two groups of animals with an articulated skeleton (arthropods and vertebrates) are the only ones that have been able to fully adapt to life on land, where the presence of legs is fundamental for movement.

The articulated appendages converted the trilobites to *all-terrain* animals, skillful in the task of moving along the surface of the substrate and also fully capable of penetrating its interior. Apparently, the first trilobites took advantage of their tremendous capacity for movement, especially when it came to foraging in the sediment for food. It is possible, moreover, that some of them also sustained themselves on algae and there were even predatory forms, although the study of the anatomy of their oral region indicates that they were limited to small prey with no skeletal parts.

The advent of mineralized skeletons constitutes one of the most enigmatic episodes in the history of life. Its appearance was almost sudden, in just a matter of a few million years, and it happened in several different animal groups almost simultaneously. As if it were a detective story, scientists ask themselves about the *motive* (that is to say, the advantage that determined their appearance) and *time* (or rather, why it happened when it did and not before or after) of the appearance of mineralized skeletons.

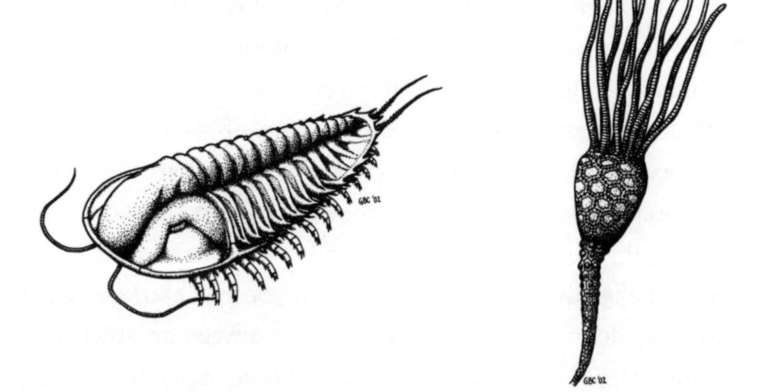

Figure 18: *Two representations of some of the first animals with a mineralized skeleton: the trilobites Olenoides and the Gogia primitive echinoderm. Both are from the Burgess Shale site.*

The motive is easy to understand. We have already said that this type of skeleton confers two great advantages: the ability of the organism to isolate itself from its surrounding medium, which includes its defense against predators, and the strength for muscular action. It is simple to understand why natural selection favored those organisms capable of being equipped with a skeleton. It is much more interesting still if the rivals, the predators or prey, also had access to this adaptation. It is conceivable that the appearance of a mineralized skeleton in some specific group increased the pressure of selection in favor of these structures and thus catalyzed their appearance in other types of animals.

But if this was so, if the motive is so evident, the problem of exactly when becomes even more complex. If it is true that possessing a mineralized skeleton is so advantageous, if its presence in a group led to its appearance in many others, why didn't this happen earlier? In the minds of paleontologists, the answer to this question is that skeletons appeared as soon as the chemical and physical conditions of the oceans permitted it. If it didn't happen sooner, it was simply because it was not yet possible.

One of the reasons cited for an explanation of why animals could not develop skeletons until the beginning of the Cambrian Period is the insufficient quantity, until then, of oxygen in the planet's atmosphere. We have previously stated that the primitive atmosphere of Earth lacked free oxygen, and this element was slowly accumulating as a consequence of the photosynthetic activity of some supremely tiny *heroines,* cyanobacteria. Even though it is relatively simple to verify the gradual increase of this gas in the terrestrial atmosphere throughout billions of years (based on the study of rocks and minerals of sedimentary origin), it is much more difficult to know precisely how much oxygen there was in the atmosphere at any given time. Different models exist, which sustain different points of view about how much oxygen there was at each point along the way regarding the increase of oxygen in the atmosphere. Some of these models conjecture that the quantity of atmospheric oxygen was not high enough to allow for the appearance of mineralized skeletons until the beginning of the Cambrian Period.

The relation between the level of oxygen in the atmosphere and the presence of mineralized skeletons lies in respiration. The amount of oxygen dissolved in seawater, a place where all animals lived during that primeval era, directly depends on how much of this gas is present in the atmosphere. If the levels of oxygen (atmospheric and, therefore, ocean-bearing) are extremely low, animals need to have an ample body surface in order to realize an efficient exchange of the gas with the external medium (to capture oxygen and yield carbon dioxide). In fact, animals use the entire corporal surface for this activity; they breathe through the skin. Under these conditions, the animal cannot cloak the surface of its body again with an isolating shield (a shell, plates, or scales), under penalty of dying of suffocation. Only when the amount of oxygen dissolved in water surpasses a critical value is it possible to cover the corporal surface with skeletal elements, transferring the mission of breathing to a particular part of the body, whose specialty is precisely that activity.

According to the defenders of the *oxygen hypothesis*, the amount of oxygen present in the atmosphere probably had not exceeded that critical value until the beginning of the Cambrian Period. This explanation is brilliant, but unfortunately, as we have already said, a consensus does not exist on the levels of oxygen in the atmosphere of that period.

Another hypothesis that has been presented in answer to the *problem of when exactly* has to do with the chemical composition of the oceans. Several investigations carried out in the last quarter of the twentieth century have made it clear that at the end of the Precambrian Epoch, the chemical composition of seawater experienced strong alterations, with marked changes in the proportions of some mineral salts, among which magnesium, calcium, and phosphorous prominently figure. These disturbances, together with the greater quantity of oxygen dissolved in the planet's oceans, could have favored the skeleton-forming biochemical processes (the cost in energy to produce and maintain a mineralized skeleton under adverse chemical conditions is prohibitive for any animal). Thus, the great quantity of calcium dissolved in the seas at the end of the Precambrian Period could have favored some organisms in developing mechanisms for precipitating this salt into the form of calcium carbonate and not dying of intoxication.

There is still another explanation, a purely biological one, for clarifying the multiplication of mineralized skeletons at the dawn of the Cambrian Period. It is the action of predators. The defenders of this point of view argue that the presence of new and more lethal predators impelled the proliferation of hard skeletons as defensive elements. This premise bumps up against two problems. In the first place, it does not satisfactorily resolve the *problem of when exactly*, since it does not explain why those conditions were not previously produced. On the other hand, and sticking with the detective terminology, animals *suspected* of being the terrible predators that this hypothesis proposes do not appear among the sediments at the beginning of the Cambrian Epoch. We have already seen that neither brachiopods, echinoderms, nor trilobites fit that profile. Nevertheless, there are traces (perforated shells and trilobite fossils with marks of having been attacked) in those ancient seas of predators that were not fossilized.

Unfortunately, this is a frequent situation in paleontology. Not all the organisms that lived in the past have been recorded in the form of fossils. To make matters worse, we don't even know how many or which of those animals are lost to us. Darwin cited the imperfection of the fossil record to explain why it did not contain all the transitional forms that his vision of evolution anticipated: it was a question of *lost links* (that is to say, imprints not preserved in the fossil record).

However, ever since Darwin's time, the work of generations of paleontologists has been tremendously increasing fossil documentation, and many of those lost links have been showing up in the strata of sedimentary rocks. At times, paleontological discoveries are broader than the discovery of some fossils (no matter how important they may be). Occasionally, paleontologists uncover exceptional sites, authentic windows on the past that allow us to know, with precision, a concrete moment in time, without the normal limitations of the fossil record. One of these marvelous places, declared a Heritage of Humanity in 1980, is the Canadian Burgess Shale site.

Today, even though one finds himself high up in the Rocky Mountains, in Yoho National Park (British Columbia), at more than 2,000 meters (6,556 feet) above sea level, the rocks and fossils of Burgess Shale are of marine origin. Approximately 515 million years ago, a prosperous community of invertebrates flourished on the sedimentary and tranquil floors of a warm and shallow sea, at the foot of a gigantic reef. Periodically, part of the sediment would slide along the steep slope of the reef toward deeper seabeds, burying the inhabitants. The animals would die and be covered by the very sediment that constituted the underwater avalanche. The sudden burial and fine texture of the sediments that entombed them allowed the soft parts of the animals' bodies to be delicately imprinted in their sedimentary casement. In this way, a complete biological community of the Middle Cambrian Period was recorded. The later convulsions of the terrestrial crust caused the sea to recede and elevated those terrains to their present-day height above sea level.

The first fossils from this site were found on the slopes of Mount Stephen in 1886. These first discoveries consisted of several specimens of trilobites and other enigmatic fossils that were baptized *Anomalocaris canadensis* (which means something like *strange prawn of Canada*), an extraordinary animal we will discuss later. At the end of the nineteenth century, the Canadian Pacific Railway constructed a railroad that ran through those same sites. The railroad workers frequently found fossils of trilobites and other animals among the rocks that they removed and called them *rock bugs*. The train facilitated access to this picturesque spot and caused an influx of tourists who continued discovering fossils. News of these discoveries reached the most important specialist in trilobites of the moment and one of the greatest American paleontologists, Charles Doolittle Walcott (1850–1927), then secretary of the Smithsonian Institute.

Walcott's scientific career was really extraordinary, if we take into account that even though he did not attend a university, and quite probably didn't even complete high school, he became president of the National Academy of Sciences of the United States, director of the United States Geological Survey, and secretary of the Smithsonian Institute. Born in Utica, New York, Walcott was left fatherless at the age of two. Obsessed since childhood with the search for fossils, upon turning twenty, Walcott took a job as a worker on the farm of William Rust, with whom he shared an interest in fossils. The collection and sale of fossils turned into a highly profitable business for Walcott and Rust, who ended up selling a collection of fossils for the equivalent of something more than 78,000 euros (US$ 84,965), by today's standards, to the eminent paleontologist Jean Louis Rodolphe Agassiz (1807–1873). What Walcott could not imagine was that this sale was going to change his perspective on fossils and the course of his life.

In September 1873, while at Harvard's Museum of Comparative Zoology helping to unpack and arrange the collection that he had just sold, Walcott was impressed by the importance that Agassiz (who would die barely two months later) gave to the study of trilobite appendages. At the time, there were no known fossil appendages, which are of primary importance in the study of these arthropods. On his return to the farm, Walcott discovered fossil remains that could definitely have included the much sought after trilobite appendages. Intrigued by the scientific importance of the discovery more than by its commercial aspect, Walcott undertook the tedious task of producing a segmental series on fossils in order to study their anatomy. In this way, in 1876, he was able to irrefutably demonstrate the existence of articulated appendages in trilobites. That same year, he began working as an assistant to the illustrious paleontologist James Hull (1811–1898), and, in 1879, he joined the United States Geological Survey (around that time he and Rust sold another collection of fossils to Agassiz's son, Alexander, for the pretty price of what is today 90,000 euros or US$ 98,000). Following that, Walcott's scientific career turned dramatic and earned him the post of director of the United States Geological Survey in 1894. By then, he was already one of the world's top specialists in Cambrian fossils.

Drawn by news of the fossil discoveries, Walcott went to the slopes of Mount Stephen in the summer of 1907. There he found new specimens, a finding that he published the following year. However, the most extraordinary discovery took place on August 30, 1909.

That day, Walcott had gone to inspect the slopes of Mount Wapta, across from Mount Stephen. When he was under the crest that joins the summits of Mount Wapta and Mount Field (really, a promontory of Mount Wapta itself), he took note of some rocks that had rolled down from the top. Compelled by his long-standing penchant for searching out fossils, Walcott, geological hammer in hand, split open some of the rocks that he had just turned over. Delicately recorded on the rock's entrails, there appeared before his eyes some wonderfully preserved fossils, including their soft tissues. Walcott tried to locate the place of origin of those rocks, but the summer was drawing to a close, and in the few remaining days on those sites, he tried to ferret out an area that showed traces of being the origin of the fossils.

Walcott returned to Mount Wapta the following summer and was able to locate, high up, the exact spot from where the loose rocks had come. He named this place Burgess Shale, because of its proximity to Mount Burgess (so named in 1866 in honor of the Canadian politician Alexander McKinnon Burgess), and began excavating (as just compensation, one of Mount Burgess's summits was christened Walcott Peak in 1996). Over the course of seven summers, Walcott and his team, which included his family, recovered around 65,000 samples.

In addition to that treasure, of incalculable scientific value, Walcott bequeathed to us an extensive scientific work on Cambrian fossils, from which he traced a third of all known fossils during that era. In the opinion of George Gaylord Simpson, one of the most important paleontologists of all time, the work of his colleagues consists of finding fossils and studying them. From that perspective, there is no doubt that the child who wandered the outskirts of Utica searching for fossils turned out to be one of the greatest paleontologists in history.

With the exception of isolated undertakings in 1930, 1966–67, and 1975, the systematic excavations in the Burgess Shale zone were resumed in 1981 under the responsibility of a team from the Royal Ontario Museum and directed by the paleontologist Desmond Collins (in which the illustrator for this book, Diego García-Bellido, a young Spanish paleontologist, took part). Thanks to the work of the team from the Royal Ontario Museum, more than a dozen new sites were discovered in the area, about thirteen miles from the spot where Walcott excavated. But discoveries of ancient sites in a state of preserved fossils similar to those of Burgess Shale have not been limited to that sector of Canadian geography, not even to the North American continent. Similar sites have been found in the United States, Australia, and Greenland.

Still, the most important site (or rather, the group of sites) was found in 1984 in the municipality of Chengjiang, in the Chinese province of Yunnan. The Chengjiang fossils are even more ancient than those of Burgess Shale; around 530 million years old, they are as well preserved as those of the Canadian site and, on occasion, more complete.

Similar to what happened with the Ediacaran fossils, the interpretation of the Burgess Shale fossils has varied since Walcott's first studies. For Walcott, all the forms represented in the Burgess Shale strata corresponded to variants, more or less primitive, of the known species of animals: the Burgess Shale fauna constituted an antecedent to present-day fauna, not anything new. However, this point of view was going to be refuted by researchers who succeeded Walcott in the study of the Burgess Shale fossils.

At the beginning of the 1970s, a team of scientists from the University of Cambridge, organized by Harry Whittington and his doctoral students, Derek Briggs and Simon Conway-Morris, reinitiated the study of the collection of Burgess Shale fossils. The analyses that were produced, over the course of fourteen years, by Whittington and his team, led them to propose a revolutionary interpretation of the Burgess Shale fossils. According to them, many of those fossils could not be attributed to any of the known large classes of animals, unless they represented animal forms that were unknown up to that moment.

Suddenly, our vision of the history of life was shaken to its core. Until then, it was thought that biodiversity in basic animal forms (that is, the number of different types) had remained constant (or had slightly increased) since the Cambrian Epoch, and what the studies of Whittington and his colleagues were suggesting was exactly the opposite: biodiversity had been much greater in those ancient seas, and from then until now, it had only decreased. This idea was picked up by the celebrated American paleontologist Stephen Jay Gould, who, in his famous book *Wonderful Life,* made it known to the general public. An arduous polemic was then initiated, which continues until our day, on the interpretation of those problematic Burgess Shale fossils. Good examples of this controversy include the cases of *Anomalocaris, Opabinia,* and *Hallucigenia,* some of the most emblematic fossils of Burgess Shale.

We have already seen that among the first fossils found in Mount Stephen were some *Anomalocaris* specimens. In fact, they were not complete samples, but rather, as was later demonstrated, alimentary appendages of a large animal. In 1911, Walcott discovered a singular fossil that he attributed to a cnidarian owing to its radial symmetry, and which he called *Peytoia nathorsti*. The first ones to realize that the *Anomalocaris* and *Peytoia* fossils corresponded, in fact, to different parts of the same animal (appendages and chewing apparatus, respectively) were Whittington and Briggs, who kept the name *Anomalocaris* for the new animal. According to these authors, this creature could not be attributed to any of the

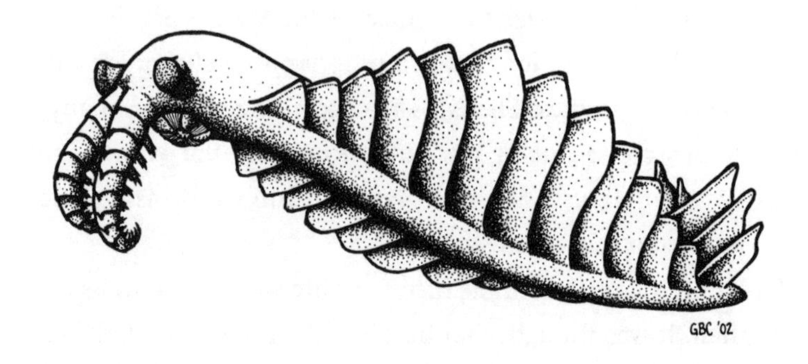

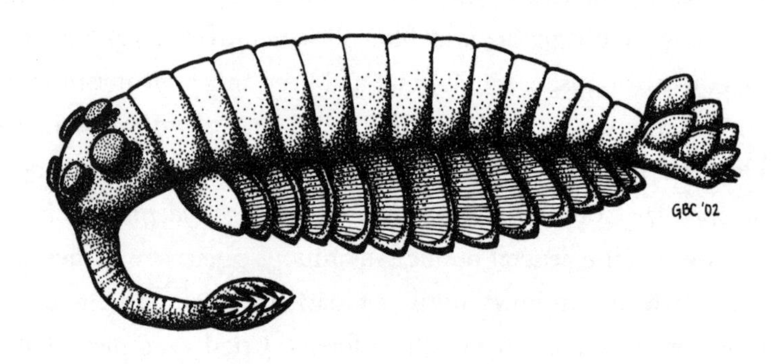

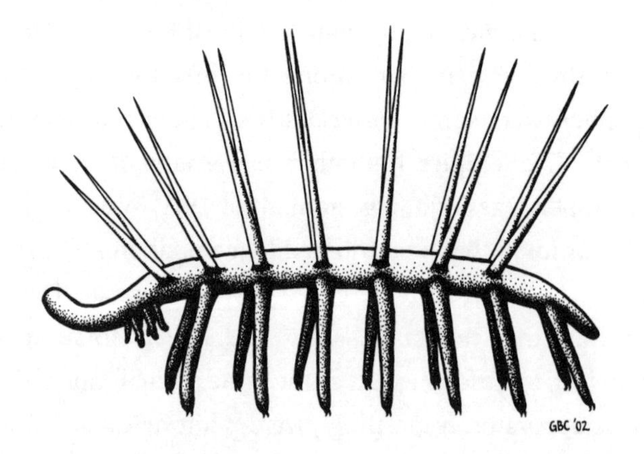

Figure 19: *Reconstructions of the most emblematic animals of Burgess Shale:* Anomalocaris canadensis *(top),* Opabinia regalis *(center), and* Hallucigenia sparsa *(bottom).*

previously known species of the animal kingdom. The excavations headed by Collins have yielded some complete specimens of this surprising animal, confirming the reconstruction made by Whittington and Briggs. Thanks to the new fossils, we know that *Anomalocaris* was the largest of all the marine inhabitants of its epoch, with a length of up to 60 centimeters (although in Chengjiang, specimens have been found that surpass a meter in length). Its body was flanked by level or flat elongations, and it had two prominent eyes and a pair of long, sturdy, articulated appendages on the sides of its mouth, which possessed a powerful chewing apparatus.

Unlike Whittington and Briggs, Collins thinks that the presence of articulated appendages in *Anomalocaris* justifies its inclusion within the arthropod groups. What they do, in fact, agree on is the idea that *Anomalocaris* was a predator, the largest in the Cambrian seas. In order to embrace *Anomalocaris* and other similar animals, all of them predators, Collins created a new class within arthropods. Owing to their carnivorous nature and given the terrifying aspect that *Anomalocaris* must have had in life, Collins christened, with a fine sense of humor, this new strain as *dinocarides*, which translated means something like *frightful prawns* (the term dinosaur means *frightful lizard*).

The most eccentric inhabitant of the Burgess Shale seas is *Opabinia regalis*. It is an animal of more modest dimensions than *Anomalocaris* (around 10 centimeters in length). As in the latter, its body is elongated and flattened expansions appear on its sides. The extraordinary thing about *Opabinia* resides in its five eyes, two paired sets and one in the center of the head, and above all, in proximity to a long appendage, similar to a trunk, located above the mouth and finished off by a crown of spiny extensions, as if it were a prehensile organ. Quite possibly, *Opabinia* used its odd proboscis for trapping its prey. Today, no consensus exists on *Opabinia's* classification. Some consider it an arthropod (of a very primitive type, such as Walcott believed, or as part of the dinocarides, such as Collins thought), while others, including Whittington and Briggs, believe it is not an authentic arthropod, but instead belongs to a new group.

One case in which their interpretation has truly reached a high degree of consensus is that of *Hallucigenia sparsa*, interpreted by Walcott (who named it *Canadia sparsa*) to be an annelid of the sea worm type. Conway-Morris produced the first detailed study of this animal, and it was he who named it *Hallucigenia sparsa*. In his opinion, *Hallucigenia* was not an annelid, as Walcott had asserted, but belonged to a new group of the animal kingdom unknown up until then. In fact, the animal's physical aspect seemed to justify the name as well as Conway-Morris's interpretation. Of modest dimensions (2 centimeters long), the body of *Hallucigenia* appears to be completed, on one of its extremities, by a protuberance. On one of its sides, it displays a double row of seven prominent spines, while on the opposite side, seven short tentacles are positioned in a single file. According to Conway-Morris's reconstruction, the animal walked on the sedimentary bottom supported by the double row of spiny extensions, while it gathered food through the seven tentacles aligned in a single row on its back.

However, this interpretation began to be put in doubt by the discovery in Chengjiang of a fossil, *Onychodictyon ferox*, similar to *Hallucigenia sparsa* and clearly assigned to the *onychophorans* species, which today is a group of sparsely numbered animals restricted to tropical forests, whose best known representation at present is the *Peripatus* genus. Generally, onychophorans are considered a type of evolutionary intermediate animal between annelids and arthropods. Along their dorsal region (the back),

Onychodictyon displays a double row of short spines, something very similar to *Hallucigenia,* which led its discoverers to maintain that, upon placing the spines on the animal's stomach, Conway-Morris had made an inverse reconstruction of *Hallucigenia.* This new interpretation of *Hallucigenia* explained the row of tentacles as locomotive appendages, similar to those of onychophorans. However, this hypothesis clashed with the fact that onychophorans do not display just one row of tentacles on their stomachs, but two (supplying these animals with *little legs*): one row on the right and one on the left.

To put an end to the doubts, Lars Ramskold, one of the discoverers of *Onychodictyon,* made a new study of the original Burgess Shale fossil of *Hallucigenia* and found the second row of tentacles, which had remained hidden in the interior of the rock that contained the fossil. Today, no one doubts that Halluc*igenia sparsa* is anything but a primitive onychophoran.

Still, there is another fossil species from Burgess Shale that deserves our attention. It is a small organism, with the appearance of a worm, that can be considered one of the first representatives of our own genus, chordates.

Ready and Ferocious

We humans form part of a propitious group within the animal kingdom, given that in the present biosphere, there are about 50,000 known species of chordates (the majority of which belongs to one of its divisions: *vertebrates*), the fourth most numerous group in the animal kingdom. Ahead of us, as far as the number of species, are mollusks, with about 100,000 species, *nematodes,* with about half of a million species, and arthropods, which greatly exceed one million species.

Included in the chordate family, besides vertebrates, is an

apparently very heterogeneous group of little sea animals that are divided into two categories: *urochordates* and *cephalochordates* (in other words, the chordate genus is composed of three subgroups: vertebrates, urochordates, and cephalochordates). Within the urochordate group, the most widely known animals are the *ascidia* (or sea urchins), which live attached to the ocean floor and are consigned to filter-feeding. Even though cephalochordates (*amphioxi* or lancelets) are not very well known in our country, they are abundant in the seas of other latitudes and are the object of exploitation on the part of human beings. Amphioxi are shaped like small fish, devoid of eyes, and live half-buried in the sediment of the ocean floor where they find their food.

If one closely looks at ascidia, amphioxi, and any vertebrate, it is difficult to spot the characteristics that those very diverse animals have in common and which cause biologists to group them in the same family. These traits, shared by all, are basically four: the presence of a *notochord*; the existence of a nervous cord that that runs the length of the dorsal region (the back) of the animal (which is technically known as the *epineurium*); in the pharynx (the throat) of the animal, there are a series of openings (*pharyngeal slits*) that are connected to its exterior; behind the anus is a muscular prolongation of the body: the *tail*. To really understand the morphology and importance of these traits, it is worth spending a little bit of time on each of them.

The notochord is something like a large center beam, located in the dorsal region, which supports the chordate's body. In urochordates and cephalochordates, this structure is a pliable rod composed of a nucleus of cells and liquid coated with a fibrous woven sheath. This same situation is found in the embryos of vertebrates (and in *myxines* or hagfish, vertebrates relatively similar to *sea lampreys*), although over the course of embryonic development, it is

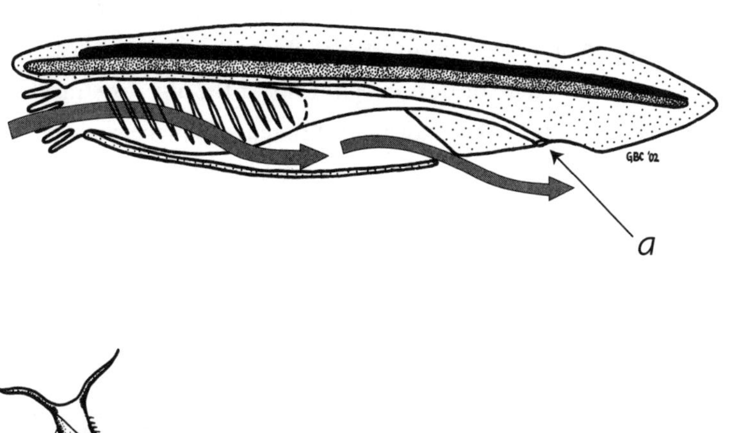

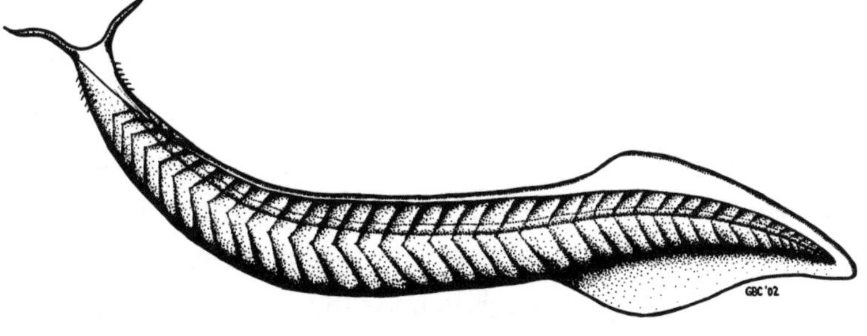

Figure 20: Sketches showing the location of the nerve cord (in black) beneath the digestive tract (in white) in an invertebrate (above), and the location of the nerve tract (black) above the digestive tract (white) in a chordate (center). In the chordate, one can also observe the existence of the notochord (perforated), the pharyngeal slits through which water exits after being ingested through the mouth (the arrows show the direction of the water), and the existence of a prolongation of the animal's body (the tail) placed behind the anus (a). Below, the chordate fossil Pikaia gracilens, *from Burgess Shale.*

gradually replaced by a series of hardened segments (vertebrae) that produce the spinal column. The notochord (or, in our case, the spinal column) not only supports the body, but also provides a muscular attachment to a powerful musculature that is responsible for the undulating movements, of the body and tail, which propel the animal.

In many animals, the brain is connected to a nerve cord that runs the length of the body in order to innervate the different organs. In nonchordate animals (those we normally group under the heading of *invertebrates*), this nerve cord runs through the ventral region of the body, beneath the digestive tract. However, in chordates, the nerve cord is found in the dorsal region, above the digestive tract (in us, it is our spinal cord). Besides this difference of placement in the body, the nerve cord of chordates and invertebrates differs in another fundamental aspect: while in the latter (invertebrates) it is a solid cord, in the former it is hollow (therefore, instead of a nerve cord, it is customarily referred to as a nerve tube). This difference is primarily derived from the different way in which this cord/tube originates during embryonic development in both.

The pharyngeal slits are frequently (and erroneously) called *bronchial slits*, in reference to the respiratory organs, which in many vertebrates, are located here. These animals force a stream of water through their mouths to the pharynx, from which it is then expelled through the pharyngeal slits into the external medium. The object of this circulation is to make the water pass through the bronchi located in these slits. Nevertheless, in the case of invertebrate chordates (and in many primitive forms of the latter), the pharyngeal slits have no connection to respiration, but rather to feeding: upon passing through the pharyngeal slits, water comes into contact with structures that filter and extract food particles from it.

If one were to ask where our pharyngeal slits are located, the answer is that these structures appear in the first stages of our embryonic development and then close. All except one. From the first pharyngeal slit of our embryos, the ducts and cavities of the external and middle ear develop. In the end, our middle ear is connected to our pharynx through a supremely delicate conduit called the *Eustachian tube* and to the exterior ear through the external auditory tube.

The muscular tail is an apparent trait of the external anatomy of chordates and constitutes a superb appendage of propulsion. In pisciform vertebrates (fish), the tail continues to perform its primitive function, but in terrestrial vertebrates, this organ has undergone multiple adaptations to the most diverse functions, even to the point of being almost completely reduced. In human beings, the tail was reduced to a group of tiny vertebrae, the *coccyx*.

Most urochordates (ascidia) only display some of these characteristics during the larval phase and lose them in the adult stage, while others (the group of *larvacea*, for example) keep them their entire lives.

Moreover, cephalochordates and we vertebrates share a characteristic that urochordates do not have: the presence of segmental muscles (*miomera*) joined to the notochord (or the spine). This type of musculature is called segmental because it is composed of a series of muscular units sequentially repeated. The best example of this pattern can be found the next time you eat fish (trout is ideal for this purpose). You will then observe that the meat of the animal comes apart in slices. Each one of them is a miomera, and the entire group forms the segmental musculature.

Finally, we vertebrates distinguish ourselves from cephalochordates in other ways. A good portion of vertebrates have a mineralized skeleton, owing to the formation of bone (even though there are many vertebrates, like myxines, sea lampreys, sharks, or skates, which do not produce bone but cartilage). This skeleton (whether bone or cartilage) has two characteristic structures. In the first place, as we have already mentioned, vertebrae (hence the name vertebrates), and in the second place (but equally important), a group of integral parts that cover the brain and principal sensory organs of the head: the cranium (many biologists prefer to use the term *craniates* to refer to the group formed by vertebrates, except for myxines which have no vertebrae). The presence of the cranium betrays another highly important characteristic of our biological group: the large development of the head and its associate organs: the brain and sensory organs (like the eyes). Last, it is also important to emphasize that we vertebrates have a heart, while cephalochordates do not.

Once we have familiarized ourselves with the different types of chordates and their characteristics, we can proceed to look at their origin, and to do that, we will return to the Burgess Shale and Chengjiang sites.

For a long time, the origin of vertebrates and chordates has been a thorn in the side of zoologists, in general, and paleontologists, in particular. At the beginning of the decade of the 1970s, the well-known fossil record of the animal kingdom's principal lineages went back to the beginning of the Cambrian Epoch (or even before, from the Precambrian epoch). However, and in contrast to that situation, chordates appeared much later in the fossil record, at the end of the Cambrian Era, a few forsaken enigmatic fossils, called *conodonts* (in fact, we refer to them as *eucondonts*), that some interpreted as the first chordates.

Conodonts are very small fossils, requiring a very powerful magnifying lens or microscope in order to be studied, and they consist of tiny skeletal pieces that remind you (some of you more than others) of teeth in vertebrates. But it is not their form, but rather their chemical composition that caused many paleontologists to link conodonts with chordates. For decades, conodonts constituted one of the greatest enigmas in paleontology and were explained as such in the lecture hall. The opinion that enjoyed the greatest support was the one that interpreted conodonts as pieces of the oral apparatus of a primitive chordate, and not having any other mineralized part, it was not preserved as a fossil. However, there were also those who thought that interpretation too perilous and without foundation.

The enigma was definitively laid to rest in 1983, when a team headed by Briggs announced the long hoped for discovery of the first complete fossil of a *conodont animal,* which they named *Clydagnathus windsorensis*. It is the fossil of an animal approximately 4 centimeters long that has a series of conodonts placed in the body's anterior region (where it was presumed they should be if, as many believed, they formed part of the animal's feeding apparatus); it also has a series of structures that have been interpreted as notochord, miomera, and tail, and it has been suggested that *Clydagnathus* has two large lateral eyes. All these characteristics clearly indicate that *Clydagnathus* (and by extension, all conodonts) is a chordate and, perhaps, a vertebrate. This last possibility is based, above all, on the presence of lateral eyes, something on which there is no consensus. Discoveries of new animals of conodonts have been going on ever since, some of them being found in posterior epochs like the Silurian and Carboniferous Periods.

Nevertheless, confirmation of conodonts as members of the chordate genus was slow in coming to the debate on the group's origin. By then, Conway-Morris had already proposed some

Burgess Shale fossils (and therefore, much more ancient than any of the conodonts) as candidates for the title of *first chordate*. These fossils had originally been placed, by Walcott, in the *Pikaia gracilens* species, which he interpreted as annelida. However, the new study effected by Conway-Morris considered one series of characteristics to be of great importance: on the tenuous impression of the animal's body, marks could be distinguished that Conway-Morris interpreted as segmentary musculature and notochord, characteristics that, as we have already seen, are typical of some chordates (in particular, cephalochordates and vertebrates).

The new interpretation of *Pikaia* was heartily embraced by the majority of researchers, and this species generally came to be considered the most ancient chordate. For some, it was the first of us. However, no one saw the *Pikaia* fossils as anything other than a primitive cephalochordate, so that the question still remained up in the air: when did the first vertebrates appear? Some were of the opinion that this position must have been occupied by the animals of the conodont genus, like *Clydagnathus*, but others were not so sure.

In the final years of the past century, discoveries of extraordinary fossils have been made that allow us to approach our own origin with greater precision. The scene of these discoveries is very far from Burgess Shale, in space as well as in time; the focus of attention has shifted to China and retraced some 15 million years to the time of the Early Cambrian Period, which is recorded on the rocks of Chengjiang. In 1955, a team of Chinese scientists, led by Chen Jun-Yuan, announced the finding of the most ancient fossil attributable to a chordate: *Yunnanozoon lividum*.

Before continuing, and owing to the complexity of the species' names that follow, it is appropriate for us to spend a moment on the question of how scientists come up with the names for the species. Naming a new species is something very serious for a scientific naturalist. Frequently, heated debates take place among scientists on what the correct name of a species is and who its discoverer was (that is, the one who named it in the first place). In light of this, some international norms have been established for determining how and under what conditions new species are to be named. This series of norms is called the *International Code of Zoological Nomenclature* (for animals), and an international committee exists to oversee its correct application. As you see, naming a new species is not a laughing matter.

This code requires that the names be Latin or Latinized and can make reference to multiple purposes, such as the place where the species was discovered, name of a person, or specific characteristics of the species named. Throughout the text, we have tried to facilitate the meaning of some of the species' names mentioned, with the goal of making the scientific *jargon* more intelligible. As many of the fossils to which we will be referring next have names that allude to specific areas of Chinese geography, it is worth our while to stop for a moment and explain. In fact, the name Chengjiang does not refer to only one site, but to a group of them, an area where you find rocks of the same age and contain the same type of fossils. We have already commented before that these sites are found in the Chinese province of Yunnan (thus the name *Yunnanozoon*: animal of Yunnan), whose capital is Kunming, the most important city in the region where the fossils of Haikou are embedded, even though there are other smaller populations, like Chengjiang (which gives the name to the group of sites) and Eraicun. Once equipped with this knowledge of Chinese toponymy, we can continue with fossils.

We said that the *Yunnanozoon lividum* species (the ashen-colored animal of Yunnan) had been proposed by Chen and his colleagues as the most ancient chordate known. This affirmation sparked a passionate polemic, since another team of researchers, headed by Shu Degan (of which Conway-Morris was a member), attributed this species to another group of animals similar to (but not exactly the same as) that of chordates: *hemichordates*. A third opinion is that of the Spanish paleontologist Patricio Domínguez, a specialist in these fauna, who maintains that *Yunnanozoon* is a very primitive vertebrate.

In open debate on the nature of *Yunnanozoon*, Shu and his colleagues announced, in 1996, the discovery of another fossil from Chengjiang and which, in their opinion, was a better candidate than *Yunnanozoon* for consideration as the most ancient chordate. The fossil in question was christened by the scientists with the delightful name of *Cataymyrus diadexus* (which means, in Greek, Chinese eel, *Cataymyrus*, of good fortune, *diadexus*). The characteristics of this fossil (segmentary musculature and notochord) allow it to be assigned, quite confidently, to the cephalochordate group. With their reference to good fortune (*diadexus*), Shu and his colleagues were expressing their desire to find more and better fossils of this type of animal.

And so it was, since barely three years later they announced the finding of two new species of fossil chordates: *Myllokunmingia fengjiaoa* (which is a mixture of Greek and Chinese vowels that means: beautiful, *fengjiaoa*, Kunming fish, *Myllokunmingia*) and *Haikouichthys ercaicunensis* (which is difficult to translate, but it means something like Haikou fish, *Haikouichthys*, found in Eraicun, *ercaicunensis*). This time Shu and his team went farther, since their interpretation of these fossils is that they are the first authentic vertebrates. Their assertion is based on, in their opinion, the certainty that in these fossils, it is possible to detect characteristics belonging to vertebrates, like cranial remains (in *Haikouichthys ercaicunensis*) and the cavity where the heart is housed (in both).

But good fortune decided to smile on Chen's team also, since barely a month after the announcement about those fossils, an article of his appeared in which they presented another new species (named *Haikouella lanceolata*, in honor once more of the city of Haikou and because of its lanceolate form), whose characteristics (the presence of a heart, lateral eyes, and developed brain) also turned it into one of the first vertebrates. If the discovery of Chen and his colleagues was unable to surpass Shu and his companions in antiquity (both are from the same strata of Chengjiang), then indeed, at least, in quality and quantity, since their discovery was not tied to a single species, but to more than three hundred, thirty of which are complete.

It is still too soon to evaluate in depth the importance and meaning of these discoveries (one as much as the other), but what really seems clear is the fact that there were already chordates at the start of the Cambrian Era. If we had to bet our money on whether some of these fossils (or all) represent the first vertebrate and to judge on the basis of the material published, we would put our chips in the *Haikouichthys ercaicunensis* basket.

There is no doubt that since the days of *The Origin of the Species*, paleontologists have earned their keep. Throughout the last century, discoveries of sites and extraordinary fossils were made that helped us to fill the tremendous void that existed in the fossil record during Darwin's time. Carrying out their expectations, those regions, whose geology was unknown, were explored, and the results obtained have surpassed all forecasts. At the same time, paleontologists have completed exhaustive studies on the new fossils that have been discovered. The moment has arrived for us to ask ourselves where we are in the old polemic on the appearance of the majority of animal lineages. Have Darwin's hopes been fulfilled, and today do we have a rich record of Precambrian animals that testifies to the idea that diversification in the animal kingdom was a gradual process that took place over a lengthy expanse of time, or has the idea been reinforced that it was a sudden, explosive event?

Even though no consensus exists, it seems that most research on the genetic material of present-day animals lines up with what Darwin proposed: diversification in the animal kingdom took place a long time ago. However, in the field of paleontology, the question does not seem so clear. Some believe now that we really know the strata at the end of the Precambrian Age, it has been demonstrated that most of the species of the animal kingdom do not have roots that go far back into time. For those who hold this opinion, fossils of the Ediacaran fauna clearly tell us that the majority of animal species of the current biosphere still did not exist at that time, and therefore, they must have originated in the short lapse of time between the end of the Ediacaran fauna and the Chengjiang Period, where we find most of these species (including cephalochordates and vertebrates) were already diversified.

Other paleontologists maintain the contrary opinion and do not believe that fossils of the Ediacaran Era can be used as an argument for denying the existence of the different lineages of animals much beyond the Cambrian phase. In their opinion, the first representatives of the different species probably lacked mineralized skeletons, which would make fossilization very difficult. Moreover, the farther we go back in time, the older the rocks, and thus it is correct to expect them to be more altered by geological processes, internal and external, whereby only a miracle would make it possible to find fossils of the first authentic animals.

As you see, a century and a half after Darwin, and following the discovery of hundreds of thousands of new fossils, there has been almost no change in the positions held by those who think that the fossil record is sufficiently complete in order to know the most ancient phases of the history of life, and those who are of the opposite opinion.

In any case, what really seems clear at this time is that when the animal kingdom burst onto the fossil record in a very decisive way, chordates, and very probably we vertebrates too, were already on the scene. The combination of a rigid dorsal axis (the notochord), well-developed musculature, and a propelling organ (the tail), displayed by our distant ancestors, allow us to know that they were among the most agile swimmers in the Cambrian seas, which would serve them, no doubt, in eluding predators as efficient as *Anomalocaris*.

At the end of the Cambrian epoch, vertebrates already had predatory forms among their ranks, like some euconodonts and, perhaps, the distant ancestors of current-day sea lampreys. During that same era or shortly thereafter, at the start of the Ordovician Period, the time had come for vertebrates to *discover* the mineralized skeleton, and they began synthesizing bones, which they used to shield themselves with thick bone plating. It was the epoch of the *ostracoderms*, or *armored fish*.

Even though the crucial adaptation in our evolutionary history was somewhat late in coming, at the end of the Ordovician Period, around 450 million years ago, the first vertebrates appeared with a jaw. Prior to that, vertebrates lacked mobile pieces in their mouths with which to process food, and this limited the type of resources they could access. Most nonjawed vertebrates could only find their food by filtering water or sediment from the bottom. This situation radically changed with the appearance of the jaw. Jawed vertebrates (technically, *gnathostomata*) had (we had) a very effective instrument for the task of capturing and swallowing prey. From that moment on, and equipped with their old adaptations that turned them into splendid swimmers, vertebrates began an intense exploitation of their new predatory ability. The oceans were filled with new and very diverse forms of vertebrates that hunted each other (and other animals) in an animated dance of predators and fish. In this context, selective pressures were emphasized that promoted the development and improvement of two more primal characteristics of vertebrates: sensory organs and the central nervous system (the brain). The need to perceive and process the greatest quantity of possible information became peremptory, and as a consequence, behaviors grew more complex at each successive stage. The process had begun that, hundreds of millions of years later, would give birth to *intelligence*.

Chapter VIII

Queen Christina's Clock

A Long Tube

In the first chordates, the central nervous system was nothing more than a hollow tube, the origin of our spinal cord, situated in the dorsal region of the body, where sensory fibers (or *afferents*) terminated and from which motor fibers (or *efferents*) exited. Among the sensory neurons, which provide information about the body (from the muscles and tendons) and its exterior (from the skin and sensory organs), and the motor neurons, which transmit impulses to the muscular fibers and glands, some neurons called *associative* or *interactive neurons* are located in the central nervous system.

The primordial function of the spinal cord is to accommodate the *reflex arc*, that is, the pathway of immediate response to a stimulus, like when we pinch a finger. But it is not just three neurons that participate in a reflex arc, since the tail ends of the neurons can be divided into several branches, which greatly increases the number of possible combinations among sensory and motor neurons through the associative neurons; in this way, a single sensory stimulus reaches many motor neurons, and a motor neuron is stimulated by many sensory neurons.

Chordates of the amphioxi type still have a simple cord as a central nervous system, with just a slight but visible widening in their anterior region; their sensory organs are barely developed. But in vertebrates, including fish, the response to information that comes from the environment is greatly expanded upon in the brain, which is located in the extreme anterior of the nervous system because this is where the sensory organs are located. The function of the brain is precisely the same as that of the spinal cord, only much more magnified: it has to do with the placement of a greater quantity of neurons among the neuron that receives the sensory stimulus and activates the muscular fiber, so that the response is more complex and also displays memory. A series of centers in the brain moves through it, gradually assuming control of behavior, which becomes less automatic each time. Ultimately, the centers associated with the functions of learning appear: memory and, in our species, consciousness.

In fish, amphibians, and reptiles (those called *inferior* vertebrates), the brain can be easily divided into three parts—*anterior* (forebrain), *middle* (midbrain) and *posterior* (hindbrain)—each one of which is associated with a sense (smell, sight, and hearing, respectively). In mammals, which are the animals with the most developed brain, the tubular structure of the brain's most ancient part—called the *stem* or *cerebral trunk*—can still be recognized. In the dorsal location[1], the three parts of the brain develop an excrescence with a *cortex* of gray matter, formed by bodies of stratified neuronal cells (arranged in layers).

The posterior encephalon (the hindbrain or *rhombencephalon*) is an important region; the sensorial fibers (or afferents), which run the length of the spinal column, and the motor fibers (or efferents), which are channeled to the muscles of the body from the neck

1. In the orientation of anatomical structures, we are modeled after a quadruped animal, in which the head is anterior (but not superior, as with humans, who are bipeds), the back is superior (and not posterior, as in humans), and the stomach is inferior (but not anterior).

down, cross it or end there. Moreover, it controls the pulmonary action in part. All the cranial nerves, except for the two (which go to the nose and eyes), start from the hindbrain or from the contiguous part of the midbrain; the twelve pairs of cranial nerves are sensory nerves, motor nerves, or both, and they belong to different regions of the head and pharynx.

The hindbrain contains the *rachidian bulb* (or *lobe*), which is the extension of the spinal cord, but larger. The nerve nuclei that reside there are essentially the extension of the four columns of gray matter that run the length of the spinal medulla on each side, two sensory columns above and two motor columns below. But located above all the nuclei is a very interesting area that is a receptor of stimuli from the ear, and in fish, from the lateral line pathway (which is a sensory system shared with the aquatic larvae of amphibians and which records vibrations in the water).

In mammals, the medulla narrows in its anterior region, forming the *protuberance* or *bridge*. The medulla and bridge belong to the stem or cerebral trunk, and the *cerebellum* develops on top of it. The cerebellum can reach a considerable size in birds and mammals (the more mobile vertebrates), and it is a fundamental organ for motor coordination and balance preservation, which is automatically carried out. The connection between the ear and hindbrain can be surprising, but it has an evolutionary explanation: in fish, the internal ear basically provides information concerning the position of the body; the information that reaches the cerebellum comes from the muscles and tendons (what are known as *proprioceptive* senses) and from the inner ear and lateral line, passing through the aforementioned acoustic lateral area of the upper part of the medulla. In mammals, impulses are also received from the cerebral cortex, passing along the bridge. In military terms, the cerebellum can be compared to an army's general staff, which serves the commanding general, informing him of the position and status of all units and, in turn, relaying orders to them from command headquarters.

The mid-encephalon (the midbrain or *mesencephalon*) is the one that has changed the least in evolution. It is associated with *inferior* vertebrates that have sight. The fibers of the optic nerves terminate in two bundles, the *optic lobes*, located on the roof of the midbrain (or *tectum*). Information also arrives from the ear, lateral line (in fish), and nose and somatic sensory regions (of the body), and motor responses are generated. The roof is the main organ of the central nervous system in fish and amphibians, their real *brain*, but in reptiles, it has to compete with a few somewhat developed cerebral hemispheres, which in birds now figure more prominently than the *tectum* in controlling bodily actions.

It can be hard to consider fish and amphibians, with barely any cerebral hemispheres to speak of, as anything other than biological robots. In mammals, the fibers of the optic nerves terminate in the cerebral hemispheres, and instead of optic lobes, there are four tubercles (called *quadrigeminals*), the two anterior ones related to ocular reflexes (visual senses are no longer sufficient); the two posterior ones are relay stations for auditory stimuli en route to the cerebral hemispheres.

Perhaps we ought to point out here that the nature of experienced sensation (auditory, visual, cognitive, smell, balance, touch, etc.) does not rely upon the fibers that transmit it, but rather upon the superior nerve center where they reside and where the impulses are converted into senses; in those centers, they change (in some way) from chemical to psychological, and subjective experience is produced, which in us can also be conscious experience.

Not all authors, beginning with the philosopher Descartes, have accepted the idea that animals *experience* anything in their central nervous systems; for them, animals could be nothing more than automatons, or animated machines with systems of motor controls that respond to external or internal stimuli (hunger, thirst, sexual drive, heat, cold, prey, predators, etc.). It is hard to find a method for verifying whether animals experience sensations and emotions, because the best formula would be to speak with them, exactly the same as when we ask someone: Does it hurt? Are you happy?

However, when Descartes claimed that animals are merely non-feeling automata, unlike man, who would be placed in a completely different category, he was ignorant of the fact of evolution. The discovery of our origin from extinct species that we could classify as *animals* changes all that. It now seems difficult for us to accept that such a chasm exists between *Homo sapiens* and all other mammals, one that strictly labels them as animated machines and us as creatures with subjective experiences. In the end, only 1 to 2 percent of our chimpanzee and gorilla genes separates us, and not much more than that from orangutans and other close primates. Our evolutionary road diverged from that of chimpanzees only 6 million years ago, which is little time compared to the 3.8 billion years that were shared. Even though in those last few million years the encephalon changed a great deal, as we will see, it is still to a large extent the same as that of all other mammals. In short, you can think, contrary to Descartes, that your cat or pet dog has subjective experiences (but there is nothing that allows you to believe that the animal in question is aware of it).

The Brain Gets a New Envelope

The anterior encephalon (or *prosencephalon*) is divided into two parts: the *telencephalon*, the most anterior region, and the *diencephalon*, the posterior area of the forebrain. Residing in the dorsal region of the diencephalon is the *epiphysis* or *pineal gland*, which in *inferior* vertebrates is sensitive to light when connected to an authentic third eye (with lens and retina). Therefore, it has the function of regulating the physiological day/night changes. In us, it secretes a hormone, *melatonin*, which also controls the circadian rhythms (the wake/dream cycle); this is why it has recently come into vogue in alleviating *jet lag*.

Thanks to Descartes, the pineal gland has a surprising role in the history of philosophy. A dualistic philosopher, Descartes believed the mind to be a nonmaterial substance and thinking (*res cogitans*) to be of a completely different nature from the body (*res externa*),

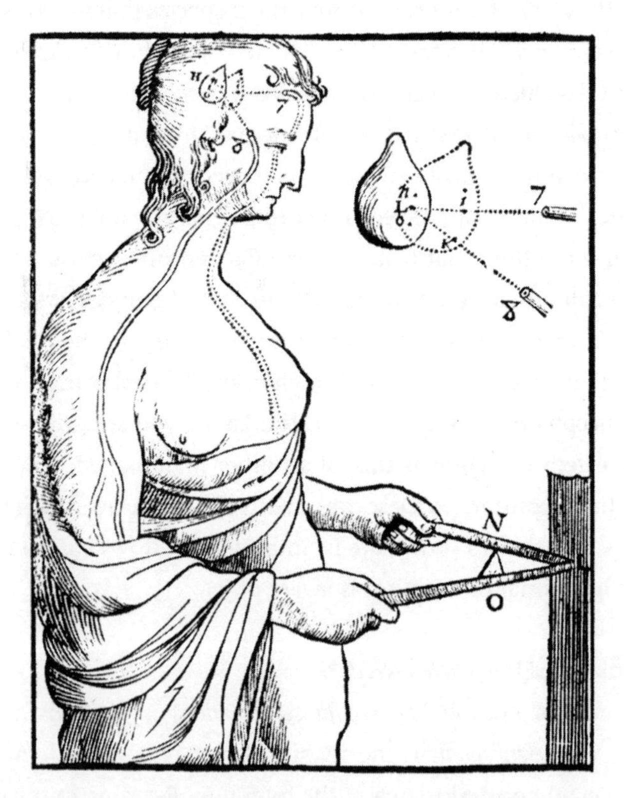

Figure 21: *In the human being, according to Descartes, the body and soul are connected to each other through the pineal gland. Sketch published in* Traité de l'homme.

which occupies space, and the body was the only thing that animals possessed. But in humans, information that passes through the senses is converted into conscious mental representations, desires which the body then obeys. But how does the mind communicate with the body? For Descartes, the answer resided in the pineal gland, that small, odd organ of the encephalon, for this was where the confluence took place: through the epiphysis, bodily stimuli generated conscious sensations in the mind, and the commands generated there were transmitted to the body and converted into locomotion.

At the base of the diencephalon lies the *hypophysis* or *pituitary gland*, which is the principal gland of internal secretion; it produces important hormones, like the one for growth or prolactin. Formed on the walls of the diencephalon is the *thalamus*, along whose dorsal area run the sensory tracts that ascend toward the cerebral hemispheres. Its importance increases as the cerebral cortex in mammals becomes more highly developed; the stimuli from the skin and ear are relayed to the dorsal thalamic nuclei, and the fibers of the optic nerve, which in all other vertebrates cross the diencephalon en route to the roof of the midbrain, have been short-circuited in the vast majority of mammals and run along the thalamus to the cerebral cortex. In short, all sensory information that reaches the cerebral cortex, except for the olfactory, goes through the thalamus.

The telencephalon is greatly developed in mammals, particularly in man, and consists of the cerebral hemispheres and *olfactory bulbs;* to be precise, it is what we call the brain. The most anterior region of the prosencephalon of vertebrates are the olfactory bulbs, the destination point for the olfactory nerves with the fibers of the nasal sensory cells. The olfactory bulbs are a relay station to the cerebral hemispheres, whose function in fish and amphibians is exclusively olfactory.

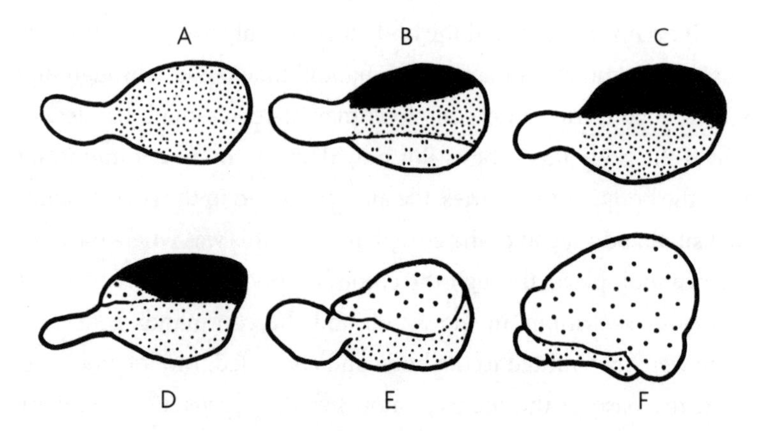

Figure 22a: *Changes in the cerebral hemispheres. (A) In the simplest type, the hemisphere is an olfactory lobule. (B) In a later stage, the amphibian, archipallium, paleopallium, and basal nuclei (above and below in the diagram) are differentiated. (C) In the first reptiles, the basal nuclei became internal. (D) In later reptiles, the neopallium is differentiated. (E) In the first mammals, the archipallium is hidden in the medial part (that is to say, the area closest to the sagittal plane or median of the body). (F) In mammals that appeared later, the neopallium continues gaining ground on the paleopallium. All the brains are represented in the same size, that is to say, on different scales (obviously it is much larger in mammals than in all the rest). Modified from Romer (1973).*

In reptiles, however, a new region on the surface of the cerebral hemisphere shows up, destined for superior cognitive functions. It is known as the *neopallium* and will end up constituting most of the hemispheres in mammals. The primitive cortex (*paleopallium*) forms what is called the olfactory lobe (or *pyriform*) and still exists in us, but is hidden in the floor of the cerebral hemispheres.

The result is that we think with one organ, the telencephalon, which emerged in connection with the sense of smell, which should not surprise us because smell is the most important of the senses in mammals, and the one that provides them additional information on the reality of the external world, even though we primates have abandoned it somewhat; on tree branches, where the life of our ancestors developed, the sense of sight is much more important than smell, as much for feeding and not getting devoured, as for calculating the precise location of the branch that awaits the final leap. We were transformed into audiovisual mammals.

Apart from the paleopallium and neopallium, there is another portion of gray matter in the hemispheres called the *archipallium* (located in the median dorsal region, that is to say, in the upper internal area of the hemispheres), which is extensive in amphibians and reptiles; in mammals it is transformed into an internal structure, the *hypothalamus*, which has a bearing on the emotional aspects of behavior. One final important element of the brain are the *basal nuclei*, which form the *striated body* of mammals and

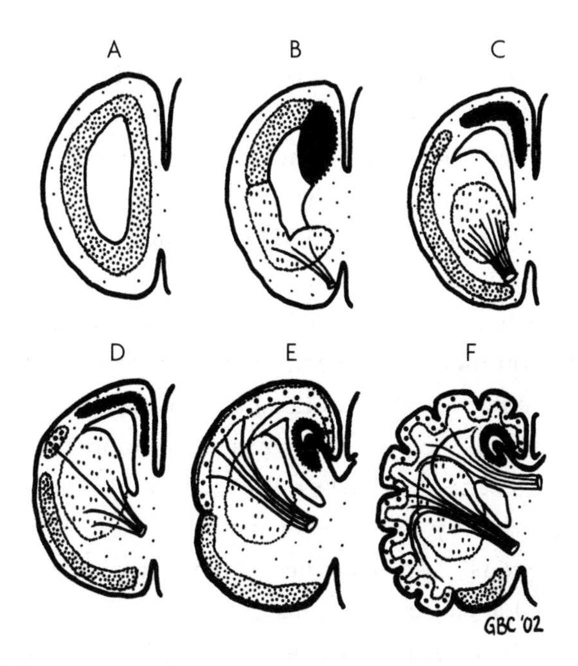

Figure 22b: *Transversal sketches of the left cerebral hemisphere in which the same elements are shown as in the anterior figure (represented with the same wefts). But here, in the first vertebrates and amphibians, one can see that the gray matter is still not on the cerebral surface, but rather is internal. On the cerebral cortex of primates, convolutions and grooves appear; the furrow that separates the paleocortex from the neocortex is called a rhinal fissure. The hollow space in the interior of the hemisphere corresponds to the cerebral ventricle. Modified from Romer (1973).*

are joined to the thalamus in both directions by bundles of fibers.

In birds, the cerebral hemispheres are larger than in reptiles, but that is mostly because of the basal nuclei, which are very important in them as centers of association. On the other hand, in mammals, the basal nuclei are located in the interior of the hemispheres; the gray cortex of the neopallium has assumed the coordinating functions that the basal nuclei, or the roof of the midbrain, exercise in the other vertebrates.

In the neocortex of the hemispheres of mammals, one observes specialized areas, called primary sensory and motor centers: one is related to visual perception, located in the occipital or posterior region, the other is the auditory zone found in the temporal lobule, and in the center, there are two regions stationed together: the anterior one is the somatic motor area that sends bundles of fibers to the body's muscles; the posterior region is the somatic sensory area that receives stimuli from the skin and muscles of the body. In these two regions, there are detailed individual representations of the human body, even though they do not preserve their real proportions: some parts occupy more surface than others.

But the entire cerebral cortex cannot be divided into specialized areas. Aside from the primary centers we mentioned, there is a large surface that has come to be called *associative*. Its function is to connect the sensory stimuli with the motor centers; generally speaking, we could say that it is the mainstay of the upper cognitive functions and intelligence in humans. It is known that the brain's anterior part (or *prefrontal*), without assigning it any single, highly specific function, is important for planning, motivation, initiative, and many other mental activities that make us human.

These unspecialized regions are precisely the ones that have broadened their surface in evolution, which, starting with the first

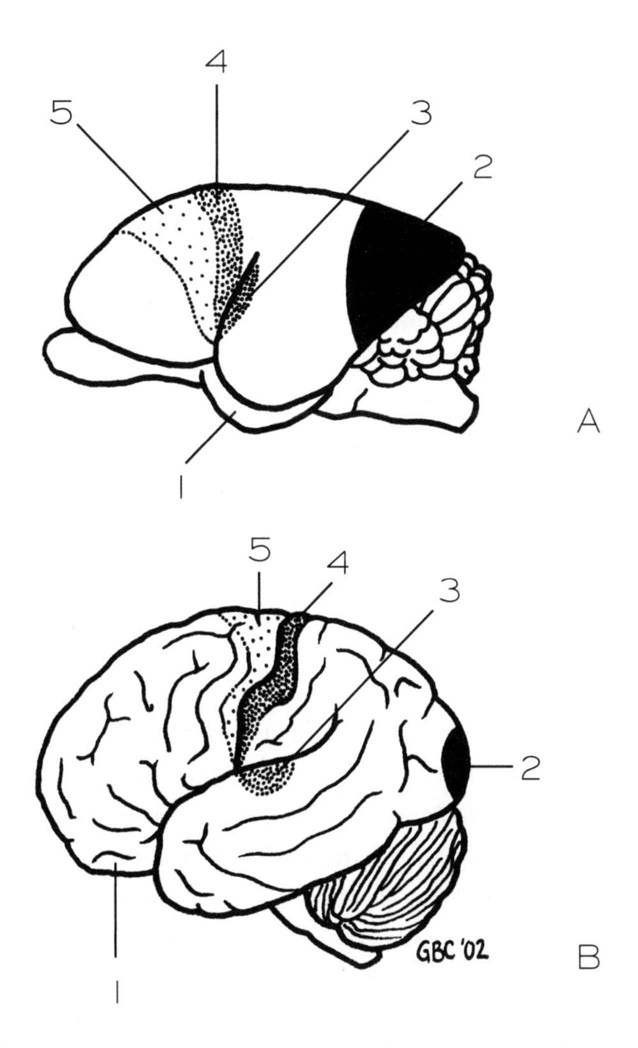

Figure 23: *In a primate with a primitive brain, like that of the lemur (A), the olfactory lobe (1) still represents an important part of the cortex, while it is not laterally visible in man (only from below). The primary cortical areas of the neocortex are: the optic area (2), auditory area (3), somatic sensory area (4), and somatic motor area (5). Among the primary areas is the cortex of association, which is proportionally much larger in our species than in any other animal. In the human brain, the somatic sensory and motor areas are separated by the fissure of Rolando, which demarcates the frontal and parietal cerebral lobules; above, the auditory area lies the Sylvan fissure, which is the superior limit of the temporal lobule in all primates.*

mammals, guided the human being through millions of years. Modern techniques of functional analysis of the brain are allowing us know what part of the cortex is activated when a person carries out a specific mental task. The exciting results that continue to be acquired sometimes confirm what had been deduced from subjects with cerebral lesions; for example, the majority of us humans have centers in the left hemisphere that are crucial to the production and comprehension of language, oral or written. On the other hand, other cognitive functions appear less localized than was first thought.

The first mammals had encephala quite similar to those of the present-day lemurs of Madagascar, though smaller. In these primates, the olfactory lobe and olfactory bulb are proportionally larger than in anthropoid primates, the group that includes platyrrhine or South American monkeys and the catarrhines of the Old World. Moreover, in lemurs, the *neocortex* (the cortex of the neopallium) is quite a bit smoother. With the expansion of the neopallium in catarrhines, the cortex grew more complex, gradually covering grooves and convolutions. Among the grooves, the *Sylvan fissure* and *central fissure* (the *fissure of Rolando*) stand out and allow the demarcation of three cerebral lobes: the *frontal*, in front of the fissure of Rolando, the *parietal*, behind, and the *temporal*, below the Sylvan fissure. To be exact, the fissure of Rolando separates the aforementioned somatic sensory (posterior) and somatic motor (frontal) areas. The *occipital lobe* occupies the posterior region of the brain, and the *perpendicular fissure* demarcates it in front. The growth in volume led to the cerebral hemispheres completely overtaking the diencephalon, the mid-encephalon, and in great part, the cerebellum, which lies directly below the occipital lobe in the human cranium.

For most of human evolution, which lasted 6 million years or a little more, there was not much progress with respect to *encephalonization*. Two and a half million years ago, our ancestors had an encephalon not much larger than that of a chimpanzee, which is not insignificant since our closest relatives in the animal kingdom stand out because of their large brain. In that epoch, the Stone Age begins, and the brain increases in size, as if culture and genes were "coevolutionizing" (in the words of Edward O. Wilson). A new enlargement takes place in something less than 2 million years ago, but as the body also grows (to current-day proportions), there is barely a relative increase in the encephalon (our early ancestors were not much bigger than chimpanzees).

If we compare the size of our brain with that of any catarrhine or platyrrhine that we would bulk up (or lower) to our own weight, it turns out that our brain is some three times heavier than that of that theoretical monkey (and some seven times heavier if we make the comparison to a standard mammal our own weight). In other words, just under 2 million years ago, the brain of our ancestors didn't even gain twice its weight. In the last half-million years, a great acceleration occurred in the rate of encephalic growth, with the result being that two very encephalized human forms appeared, or to be more exact, became highly telencephalized, which is to say, possessing enormous cerebral hemispheres; those two species of humans are the Neanderthals (extinct barely thirty thousand years ago) and us. The most anterior part of the central nervous system, destined at one time for identifying molecules in water or air, led us to technology and consciousness.

With this brief description of the anatomy of the central nervous system, especially of the encephalon, we provide ourselves with an idea of how evolution endowed vertebrates with the ability to manage their behavior, and in the case of man, to do it consciously. But there is a different way to approach this same problem that will be presented in the next pages of this chapter and the one after that; in other words, investigating the evolution of behavior and the laws that govern it.

Behavior As an Adaptation

We have already said that in *Summa Theologica*, Saint Thomas Aquinas developed five ways for demonstrating the existence of God. The fifth is the one that interests us the most, because it refers to the internal finality that exists in nature: "We see, in effect, that things that lack knowledge, like corporal bodies, work toward an end, as is proved by observing that they always, or almost always, work in the same way to acquire what best suits them; thereby, one understands that they do not achieve their end working by chance but rather with purpose. However, that which lacks knowledge does not move toward an end unless someone who has understanding and knowledge guides it, in the same way that the archer aims his arrow. Therefore, an intelligent being exists who directs all natural things to their end, and we shall call him God."

Physiological structures or functions do not merely have finality in animals for the sake of finality, since we can ask ourselves what purpose is served, the same as when considering a beak, tooth, wing, or fang as when contemplating an internal organ like the kidney or brain. In fact, the biologist's job, as defined by Daniel Dennett, consists of conducting an exercise in inverse engineering. Just as in industrial or military espionage, the apparatus designed by a rival company or enemy force is taken apart in order to see how it functions and what purpose its parts serve, biologists *deconstruct* organisms in order to be able to understand them, at the structural level as well as at the functional level. In other words, they walk backward on the road of evolution. At any rate, we should be careful in this investigation, because not all the *pieces* of living organisms serve a purpose; not all are adaptations. The reason for this is that organisms are not perfect systems hot off the drawing board of an engineering firm. As we have already stated before, selection reminds us more of the work of an amateur do-it-yourselfer than that of an engineer, as Nobel Prize-winner François Jacob points out to us. The light bulb is not a descendent of the candle, nor the jet of propeller aircraft, because engineers create new designs without relying at all upon previous ones. On the other hand, natural selection acts by modifying what came before in order to produce new functions in old, redesigned organs; there are, therefore, many aspects of organisms that can only be understood by knowing their history. In this task of inverse engineering, the isolated and static constituent elements, which are the object of the morphological sciences, are not the sole interest, for there is also the question of how they function in living organisms, and this is the mission of physiology. Therefore, the biologist has to try to study complex biological *machinery* without interfering with its operation. This investigation is carried out on several levels of organization: those dealing with structures, cells, and molecules.

But most animals are active and move about, and their behaviors also produce benefits, for they too are endowed with a purpose; and in light of this, we can also ask ourselves, what purpose do they serve? In their multiple variants, all behaviors (just like corporal organs) are aimed at assuring survival of the individual and that of his descendants, but in order to achieve it, they carry out many different functions, especially in those groups with a well-developed central nervous system. Behavior is also partially mapped out in the genes, and it is teleonomic, that is to say, adaptive and a product of evolution. If it were not so, no animal would be able to survive in an autonomous form.

Animal behavior can be broken down into a series of categories. Any animal that moves autonomously (unlike the car, tractor, or remote-controlled space shuttle, which, being as complex as they are, need someone to handle them) has to be sensitive to the information that comes from the medium in which it operates. At a minimum, it will need to distinguish the favorable environment from the unfavorable one and make use of mechanical means of response in order to persevere/advance (in the first-case scenario) or, on the other hand, to flee (in the second-case scenario). Also, any form of mobile life (that is to say, that which moves on its own accord and is not passively operated) will need this fundamental mechanism in order to explore space, evaluate it, and consequently take action (*make decisions*), since animals are not limited simply to observation.

Thus, animals (and humans too) are equipped at birth with some knowledge of the external world essential to their survival. It is not true, as empirical philosophers have asserted, that we come into the world as clean slates, tabula rasa, without any information about the world. According to this philosophy, all knowledge proceeds from the senses (from experience). Many scientists also believed that man was the only animal species that lacked instincts, except for some fundamental ones at the onset of life, especially those having to do with breast-feeding.

To the contrary, animals and man are born with knowledge. How do they get it, and how is the information stored? Information is warehoused in the genetic system, the *genomes,* which according to François Jacob is one of the three memory systems with which animals are equipped (the other two are the immune system and central nervous system). Thus in reality, when the animal confronts a stimulus for the first time in his life, it could be said that it *recognize*s the said stimulus and acts accordingly (if it were otherwise, *recognition* would serve no purpose). The genome is the species' memory, because it is a consequence of the vicissitudes through which its ancestors have passed before it, though not by way of a direct line of inherited learning, as Lamarck believed, but rather by following Darwin's route, the road of natural selection.

The genome, in effect, stores information, which has been building up through the slow mechanism of natural selection, about the external world. It is not, strictly speaking, a true process of learning, because it is not produced during the individual's life and is not based on experience, but rather the individual is already born with or without knowledge. In fact, the usefulness of the knowledge in question cannot be evaluated until the following generation. All of this seems very complicated, but it isn't. It just means that modifications in the structures and behaviors of living organisms appear as a consequence of random mutations, because of copy errors in the transmission of genes between generations. The variant favorable genetic codes are, by definition, those that have the subjects that carry them (the *mutants*) reproduce more. Therefore, until the second generation (and those that follow), it will not be known whether the mutation has or has not been a success. That reasoning is as valid for structures as it is for behaviors.

What is now important is to underscore two things: first, that individuals have innate mechanisms (they are born with them) of recognition of signs (stimuli), and second, that individual experience is not transmitted to descendants via genes.

Is There Anyone Out There?

In the simplest possible case, the animal only perceives and reacts to a specific type of stimulus (whether temperature, light, humidity, acidity, etc.): a manifest signal and an explicit response. However, animals equipped with a well-developed central nervous system (in particular, some mollusks like octopuses, arthropods, and vertebrates) are capable of reacting to many types of different signals and with a wide variety of behaviors. To do that, they make use of sensory organs that provide them with a very rich (very complex) vision of the world. How are those sensory organs arranged so that in each case, the animal selects the appropriate stimulus and responds in a coherent manner (which we would call *logical*)? It could be said that it comes with experience (as empiricists would claim), but many observations have been made that demonstrate that the stimulus-response association is often innate. We would also equip a system with programs of this type that we would attempt to turn into an automaton (and we wouldn't want it to be destroyed before sufficient experience was gathered).

Thus, a duckling hatched by a hen, as soon as it breaks open its shell, will run straight to water to the desperation of the adoptive mother, and it will swim, dabble for food in the mud, and preen its feathers like any normal duckling incubated by a mother duck. The example set by his adoptive brothers, in pecking at the grain around the mother hen, will do him no good. Those fixed behaviors, anterior to any experience, are not exclusive to animals, which to us could appear *inferior* (or to have simpler central nervous systems), such as birds. Among mammals, behaviors are also fixed. In the autumn, a squirrel, having sated itself, will store nuts that it will later find after burying them at the foot of a tree. We can conduct the experiment of raising a squirrel in a cage, under conditions of total isolation, and put it on a liquid diet. The first time we give it a nut, it will try to scratch and bury the fruit under a slat in the cage. Similarly, we could give any number of examples of innate behavior. Some songbirds are born already versed in all the melodies sung by an adult, as has been proven in experiments using isolation; ducks do not need to learn the rituals of courtship, which spontaneously manifest themselves when the animal reaches sexual maturity.

Ethologists (scientists who study animal behavior) have indicated that it is necessary to acknowledge the existence of an apparatus, a physiological mechanism that filters the innumerable stimuli that reach the organism from the external world. This device is called the *innate releasing mechanism* (IRM), and its function consists of releasing a fixed standard of behavior in the presence of a key stimulus (assuming the simplest) or of a complex combination of stimuli that reach it through multiple channels.

In fact, rather than releasing a behavior, it would be more accurate to speak of it as triggering a behavior, because ethologists believe that behaviors directly originate from the central nervous system in an *endogenous* (spontaneous) way, even though they are normally inhibited. Only when specific stimuli are actualized in a situation, does the IRM unchain those repressed components. The selectivity of the IRM resides, according to experiments, as much in the sensory organs as it does in the central nervous system.

By behavior, in its deepest physiological sense, it is understood here to mean a rigidly fixed sequence of signals, an instinctive movement. Animals that possess a hardened skeleton, vertebrates that is, with an internal bone skeleton, and arthropods, with an external chitin skeleton, are very limited in their movements, which are also limited in number because these organisms are not made of rubber: it is very difficult to coordinate the ten claws of a crab, to direct the flight of a bird or the gallop of a horse, and animals do not have to think at each step of the way what they are going to do next (our cars too, despite their incredible benefits of speed and transportation of cargo, are very limited in their possibilities of movement, as we discover every time we have to parallel park in a tight space). Therefore, behavior is structured through a chain of signals, which are fixed and innate (they are genetically programmed), and were acquired throughout evolution. These chains are called hereditary formulations or instinctive movements. The most important ethologist of all, Nobel Prize-winner Konrad Lorenz (1903–1989), considered the discovery of hereditary coordination, the real *anatomy of behavior,* to be the cornerstone of the entire theoretical structure of ethology.

An important consequence in all this is that he debunks empirical philosophy as much as he does idealistic philosophy. Idealism proclaims that all reality is basically subjective, that is to say, it is a product of our minds, and there is no way for us to know what it is that is *out there*. Empiricism is its rival: for empiricists, experience is the only source of knowledge; all true wisdom, as a consequence, is a result of experience.

That vexing doubt about what is true and what is illusion is what envelops us, and in our own experience, it is a recurring theme in literature and film, but no one dealt with it better than don Pedro Calderón de la Barca (1600–1681) in his profound drama *Life Is a Dream*. In a few very famous verses, the distraught Segismundo is confused and doesn't know when he is awake and when he is dreaming, that is to say, if what he sees is in his mind or outside it:

> What is life? A frenzy.
> What is life? An illusion,
> a shadow, a fiction,
> and the greatest good is small,
> and all of life is a dream,
> and dreams are but that, dreams.

However, the theory of evolution holds that the sensory organs and nervous system are adapted to the external subjective reality, the one that exists *outside* us (we see with our eyes as well as with the brain, which processes the information that, in the form of nervous impulses, reaches it from the retina by way of the optic nerve). Natural selection is provided for organisms to make use of methods that capture external information, if not all, at least what is relevant to the individual's survival. The IRM of every species then selects a series of signals that have to exist by necessity (objectively), because if it were otherwise, the patterns of behavior that it sets free would be devoid of any meaning and would lead to disaster. As Segismundo puts it:

> What if perhaps I am dreaming,
> even though I see that I am awake?
> But I am not dreaming, for I touch and believe
> what I have been and what I am.

Thus the old philosophical debate between idealism and empiricism overcomes its impasse thanks to the theory of evolution: each species has its own *vision of the world*, which depends on a few biological mechanisms with which it is born and have to be anterior to all experience, precisely so that any experience will be possible. This biological apparatus for absorbing objective reality constitutes the *lenses* that define the way we *see* the world, and they closely correspond, as Konrad Lorenz says, to the formation of the very important German philosopher Immanuel Kant (1724–1804). This philosopher thought that human knowledge was the result of an active mind that converted sensory experiences (*intuitions*, as he called them) into ideas; to achieve that, the mind relied upon a series of a priori forms and categories (that is to say, previous to experience), like those of space and time (the two *forms of intuition*), quantity, quality, relationship, and form (the *categories*), that permitted sensory information to be processed (to give form to what is reported).

In the words of François Jacob, "If the representation that the monkey makes of the branch onto which it wishes to leap had nothing to do with reality, that would be the end of monkeys. And if the same thing happened to us, we would no longer be here to talk about it. Perceiving some aspects of reality is a biological necessity, and just some of them, for it is evident that our perception of the external world passes through a massive filtering system."

Ethologists carry out many experiments with animals simulating key stimuli and combining them in different ways: sometimes their effects accumulate, other times they counteract one another. By raising animals under conditions of isolation, researchers assure themselves that the reactions are innate, always produced before the first appearance of the stimulus in the animal's life. An even more intense reaction from the animal can be achieved by exaggerating the stimulus, that is to say, with an artificial stimulus more powerful than the natural one.

The male robin is very territorial, and if we stick a stuffed bird into his territory, he'll immediately start pecking at it. However, it doesn't take an entire bird to cause the male robin to unleash his attack, but rather the aggressive behavior of the robin (the lord of the territory) will take the very same form before a plume of red feathers, and by contrast, there will be no attack if the stuffed animal is a young bird without the characteristic red feathers of the adult male (which gives name to the species) on its breast. It could be said that the object of the aggression is not strictly the rival bird, but in effect, the red breast feathers. They, and not the beak, for example, incite the aggression. The red feathers are a *releasing* element.

Male stickleback fish are also very territorial. The male is distinguished from the female because he has a red underbelly, while the female's is swollen. In laboratory experiments, the males did not attack male stickleback molds (decoys) that were totally perfect replicas except for the absence of a red underbelly, while they did go after almost any decoy, no matter how coarsely made, with a red underbelly, even though it lacked fins and a tail; but if the replica is inverted and given a red dorsal region, the male's aggression is no longer triggered. On the contrary, a very basic replica, but with a swollen belly, also incites the male and courtship begins.

Speaking of supernormal stimuli, it can be proven that some birds prefer to incubate an egg four times larger than their own, despite not being able to sit on it in any way because of its enormous size.

Without any doubt, innate releasing mechanisms also act in man before specific *releasing* stimuli that are perfectly synthesized with them, and this is a fact that industry knows perfectly well. Animals, which are used as toys for children, the same as the characters in comic strips, have baby-size proportions: large heads in relation to the body, short snouts and small ears, big eyes, rounded belly, etc. These figures awaken sympathy in us for the same reason that pups, kittens, and our own children do: because we are programmed to love and protect them, in the same way that the long beaks, big ears, and fangs of the hairy *Big Bad Wolf* transform him into the bad guy in children's stories.

We normally perform a very simple experiment in class (taken from Konrad Lorenz) that illustrates up to what point very simple stimuli are immediately recognized (instinctively, we could say). If we draw a circle that contains two dots and below it an arc open on top, everybody interprets it as a smiling face, but if the arc opens downward, the *face* then looks sad. Through association with expressions of the human face, the camel appears haughty and the eagle noble. Lorenz has also denounced how publicity continues to make tobacco, a lethal drug, attractive to us, by associating it with the image of youth and vitality, or even, paradoxically, with health.

Key stimuli can come from individual entities of other species, that are their predators or prey, and also from the same species, because animals sexually reproduce and need to have at least one other model of the opposite sex at the time of procreation. In those animals that look after offspring and in social animals, the need to filter the stimuli that come from all the others is very important. In this way, organs develop that function like signal transmitters (optical, acoustical, or olfactory), whereby those signals are retrieved by other individuals; the behavior itself is a signal. A problem (rather: a challenge) for the theory of evolution is the fact that these associations have come to be established between transmitters and receptors: the female grouse that responds to the strutting of the male; the parents who feed the chick that opens its beak and shows them its red throat; the robin that attacks the male intruder; the wolf that is the leader of the pack and reins in his fury when a subordinate puts his tail between his legs and lowers his ears or, faced with the threat of a dispute, offers him his neck or rolls over on his back out of naked fear; the tota monkey that climbs the tree when another member of the group sounds the alarm for: *look out below, a snake*; the male chimpanzee that recognizes in the female's red perineal swelling her receptivity to copulation, etc. The specific behavioral response would not have been produced (or at least not with the same intensity) if the male grouse hadn't shown off, if the chicks didn't have a red throat or the adult male robin a red breast, if the cowed wolf hadn't submissively offered his neck, if the monkey's acoustical signal had been that of *look out, an eagle,* or if the female chimpanzee hadn't displayed the conspicuous perineal swelling. What came first, the chicken (receptor) or egg (signal transmitter)?

It is reasonable to think that every relatively complex living organism, whether of this planet or any other, is equipped with an IRM and hereditary formulations. Even an automaton, which we might propose to build, would have to have selectivity when presented with stimuli and a few fixed patterns for movement of its structures, which we can imagine as more or less rigid and articulated. Like arthropods and vertebrates, the mobile systems (robots) that we humans build also have limited movement, especially if they are to be more useful to us.

But if we do not add another single mechanism, the system will remain static until a key stimulus crosses its paths to unleash a reaction. If what we want is for the system to do something, then we will have to tell it: *Hurry up, get moving, and go find the stimulus!*

Pulsations: Drives and Instincts

Animals are, to be sure, driven, *from within,* to search for the stimuli that allow the unchaining of behavior that has not been exercised for some time. Some internal mechanisms connected with drives are well known: hunger, thirst, hormones. The organism has internal sensors for recognizing that its liquid or glucose levels are low or for detecting an elevated rate of certain hormones in the blood. Impelled by these stimuli, animals become restless and tense, and the tension does not abate until the drive has been satisfied. In fact, as time passes, the intensity of the necessary stimulus for unchaining the pattern (the so-called *threshold*) continually diminishes (the low threshold), until the pattern switches itself off once the stimulus has disappeared, as if the animal had imagined it. This animal drive, which leads it to actively search for the stimulus, is called *appetite behavior*. Moreover, there are priorities among the instincts, according to the situation, when any of the following can prevail: the need to drink or eat, the need to court or attack, or to protect one's young or flee.

It is interesting to observe that each behavior awakens in due course, maturing as if it were one more organ: there are patterns of infantile behavior and others that only belong to adults, like those relating to territory, courtship, looking after one's children, etc.

But moreover, ethologists discovered something surprising: an internal source of behavioral drive seems to exist that probably took root directly in groups of motor cells in the central nervous system that accumulate tension (in the form of *motor restlessness*) until the activity, coordinated by those cells, is carried to completion. In effect, many of the behaviors that animals exhibit (searching for food, hiding and storing it, constructing a burrow or nest, courtship and mating, taking care of the young, defending territory, etc., plus social behaviors) could function in this manner. The most dramatic way to demonstrate this behavior is recognizing that predator animals that live in captivity periodically need to satisfy their need to hunt, even though they never lack for food, and so they *desperately* search for some prey, whether real or imaginary.

A starling that Konrad Lorenz raised and which was well fed, though it never had the opportunity to hunt, went on short flights from time to time, exhibiting the exercise of capturing an insect; afterward, it would return to its perch, where it acted as if it had killed the insect and swallowed it, and then once again, settled into a tranquil state.

There's still more. To hunt a rat, the cat crouches, tracks its prey, and then attacks and kills it, and each one of these activities is to a certain degree independent, although normally they develop interconnected, so that if the cat spends a long time without exercising any of those instinctive actions, an *appetite behavior* will develop, consisting of the search for a specific releasing stimulus for such an act: to crouch, pursue, kill, or whatever; the same rat can serve to trigger each of those separate behaviors. That production of excitement in the central nervous system that generates the need for movement seems different depending on the life habits of each of the species, from which one would deduce that it is a product adaptation of evolution. For example, lions, which normally do not need much time to chase down their prey in nature, are quite tranquil in captivity, while wolves, foxes, *mustelidae* of the marten type, or weasels, which need to be very active in the search for food, show a great need to pace back and forth in a zoo, even though they are not hungry.

Looking at it the right way, this is a very efficient way of programming an animal's behavior and assuring its survival and that of its genes. Instead of explaining, in minute detail, everything to it that it needs to do throughout its existence, its genes program the animal to search out certain, very specific stimuli, which will release the appropriate behaviors for living and reproducing. Moreover, each one of these behaviors has its own rhythm, such as having to eat several times a day but only needing to build a nest once a year. Similarly, we would also program an automaton that we wanted to perform a certain task: the automaton is a teleological object that has no aims of its own (except those of the one who built it); therefore, we could simply tell it to look for a certain signal, and once it has found it, to set in motion the programmed task, and for it to do this, let's say, twice a month; meanwhile, it could seek out other signals and carry out other activities that we are also interested in having it perform (such as storing energy for future tasks, for example).

• • •

The Problem of Infinite Regression

The fundamental difference between the automaton and human being lies not so much in their activities and the type of mechanism that sets them in motion, but the fact that automatons do not replicate themselves, and therefore do not evolve. But taking our imagination a little further, we could devise an automaton that would be programmed to gather materials from its surroundings and assemble copies of itself. In this way, we would multiply the number of systems in circulation on Earth, and this artifice could be useful to us were we to send them to another planet, where it would be impossible for us to build replicas. Moreover, the new automatons would replace the old, worn-out ones that would be operating less efficiently each time, until they finally became inoperative and *died*.

But in this daring hypothesis, we would have to equip the automaton with a means for transferring to its replica the necessary information for creating another copy. Otherwise, the chain would be broken with the first *descendant*. If, in the process of copying that information (the program for generating replicas), there existed the possibility of making mistakes, we would have something like a mutation.

Given the variations of automatons that would arise because of *mutation* along the way, natural selection could come into play, since having reached a certain level of multiplication in the number of individual automatons, perhaps there would be no way for all of them to acquire the necessary energy to function or the materials needed to produce more replicas. A competition would then ensue for resources for which not all the variants would be equally equipped, and in the end, some would prevail over others.

We would then have all the necessary requisites for an evolution of those automatons. Since the proposed model is of asexual reproduction, it is then logical to ask what purpose is served by sexual reproduction, which requires two individuals to agree on producing a third. The answer is that it serves no purpose, in the sense that sexual reproduction does not pursue any objective; it simply happens that evolutionary lineages that are reproduced sexually, at least on the planet Earth, have had more success in the long run than those that reproduce asexually, even though there are both types of organisms. To explain this fact, or to try to, would take us far afield, but we know that sexual reproduction produces more variation than asexual reproduction as well as variation in the primary material used by natural selection. The more variations, the more possibilities natural selection has for producing new designs that involve important improvements.

The automatons described here would acquire from the environment the energy and materials required of them and could, in comparison to organisms, like plants, be called *autotrophs*. But many different types of automatons could crop up that obtained their energy and materials from other automatons, from which they would snatch *life* (would dismantle it), at which point we could consider them to be *heterotrophs,* like animals.

Some authors of science fiction promote the idea that self-propagating automatons will be the only means by which we successfully explore, and exploit, the rest of the universe, given the enormous distances that exist compared to the brevity of each human generation. But still, this is an argument against the existence of any other forms of intelligent life, like ours or superior to it, in our galaxy or beyond. If they do exist, how come their self-propagating automatons have not yet reached us? Perhaps it is only a question of time, although, on the other hand, interstellar distances are mind-boggling. It is true that there are billions of galaxies where life could be sustained, but most of the universe is desolate, with terrifying chasms of isolation from one solar system to the next. Galaxies have diameters of tens of thousands of light-years and are normally grouped into clusters with distances of a million or more light-years between them. Separation between the clusters is many times greater.

The *Voyager* I spacecraft has already gone into outer space, outside our solar system, but its velocity is limited (17.24 kilometers per second). At that rate of speed, it will take a spacecraft some 600 million years to reach even our planet from the center of the Galaxy, situated some 33,000 light-years away (the diameter of the Milky Way is 100,000 light-years). If in the course of its flight, it made a stop somewhere, the trip would be even longer. Complex organisms, made up of many cells, began emerging 600 million years ago on Earth. Perhaps evolution played out more quickly on other planets, from other solar systems. An intelligence superior to ours could shorten the trip, of course, or maybe live closer, for example, in the triple star system of Alpha Centauri: the closest of the three to the sun, called Proxima Centauri, is only 4.3 light-years away (it would take *Voyager* I some 77,500 years to reach it). It is now known that there are, besides the sun, other stars with planets tracing orbits around them in our galaxy. But, to sum up, the fact is that even through the use of self-replicating automatons, we had no proof that there is *someone else* out there.

We have discussed the ability of self-replicating (or self-propagating) automatons created by man in a very cursory manner, as if it were a simple topic. The fact remains that we still have not managed to produce any that stand out.

The naturalist philosopher René Descartes was a man of very broad learning. He spent his last four months in Stockholm, where he died of pneumonia in 1650. He had been invited there by Queen Christina of Sweden (1626–1689), a historical personage who had given people much to talk about. The captain of the ship who took him to Stockholm in September of 1649 was astonished at the great philosopher's nautical knowledge. As we have seen, Descartes thought that animals were mere automatons, and it is said that one day Queen Christina showed him a clock and told him: "Make it reproduce itself." The fact is that, in effect, the ability for autoreplication is an exclusive attribute of life, which has still been impossible for us to imitate.

The problem is that for an automaton to construct another just like itself, it needs a blueprint. An engineering or architectural plan is a blueprint of a system or building, and without that schematic model of the desired outcome, there is no way for the assemblers to begin building the system or house in question, because they wouldn't know what is required of them. The fact is, robots have replaced humans on the assembly lines of automobile factories and on other complex machinery, but the robotic arms that perform certain tasks are responding to preprogramming; the vehicle that is produced does not contain an *autodescription* (an *internal blueprint*) anywhere in its parts, and therefore, it can't reproduce itself (what more do we want, for cars to multiple in our garages like horses in stables!). We could also program robots to make other robots, but the latter would be incapable of doing anything if we did not reprogram them.

The difficulty in building autoreplicating automatons lies precisely in this: there is no internal blueprint. How is it possible for a system to have a blueprint of itself that it then passes along to another machine, so that it in turn produces a third likeness? The system's internal blueprint, in turn, would have to include another blueprint, to provide details of its parts and an explanation of the design. It would be tantamount to using plans for building a machine that has an internal blueprint of itself. The assemblers could easily follow the instructions for building it. These imaginary assemblers have no clue about what is to be done and will need instructions, but since the sister machine contains its own blueprint (its own plans), it would be possible to produce a third likeness based on that blueprint. However, after the granddaughter system, there would be no way to produce a great granddaughter system, because the latter no longer contains an internal blueprint of itself. For each generation we add, another blueprint in the initial system is required. This logical difficulty is known as the problem of *infinite regression*.

Here's an example. Let us suppose that an architect wants workers to build a studio just like his down to every detail. He will then draw up some plans that will provide the assembler a description of what he wants. But once the new studio is built, no one will be able to replicate it using that plan, because it does not have an architect inside to draw up the design. The solution would have been for the architect to have drawn up a blueprint of his studio that contained every detail, including a blueprint of the studio itself. The assemblers would then construct a new studio that would have tables, chairs, lamps . . . and a blueprint inside, which could then be the basis for a third studio, but the chain would end there.

The famous mathematician John von Neumann (1903–1957) came up with the solution to this complicated matter; it consisted of auto-replication having two functions. On the one hand, it would provide instructions for reproducing a system, meaning, one that could be interpreted or translated. On the other hand, it could be copied without interpretation, that is to say, literally transcribed. Not long afterward, in 1953, James Watson and Francis Crick discovered the structure of the molecule of biological inheritance, DNA, which has those two properties. To produce proteins, the DNA in our cells is transcribed into the messenger RNA (mRNA), and in turn, the mRNA is transformed into proteins.

Moreover, when *mitosis* or cellular division is produced, the DNA is duplicated. The mistakes copied into the replication are precisely the mutations that make evolution possible.

With this type of bivalent internal blueprint, it will one day be possible to efficiently produce self-replicating systems (robots), and one has to ask if at that point we won't have to confront a mortal enemy. Will our species one day be at war with legions of mutant robots? Will our systems turn against us? But instead of becoming frightened by the prospect of our artificial offspring posing a threat to us at some time in the future, we would be much better off to concern ourselves with what we humans are capable of producing.

The Noble Savage

The existence of spontaneous or innate instincts is a hypothesis of ethologists that would not have sprung up outside the scientific community if it weren't for the fact that there is one instinct in particular among animal drives whose mere mention makes our hair stand on end: aggression. If the hypothesis concerning drives is correct, then aggression is not a mere consequence of environment, but rather an innate instinct in animals, including (at least, in the beginning) man. Consequently, aggressive behavior would not be just the result of a repressive and ineffectual education, as some philosophers, psychologists, pedagogues, or sociologists suggest (and many politicians proclaim, since the contrary hypothesis, that men are not exactly angels, is not a popular one).

Even admitting that some innate patterns of aggression exist, which are only released in the face of specific signals as if they were reflex actions, would be reason enough for eliminating any causative stimuli from our environment in order to suppress future aggressive behavior on the part of humanity. However, these very optimistic theories have always collided head-on with the harsh reality of life: if it is true that there is no aggressiveness whatsoever in human nature, why is it so hard for us to eradicate it in practice? Why is there a Nobel Prize for peace? Would eliminating the failure of education be enough to solve it? Would not giving children everything for which they ask then make them more tolerant when confronted with the inevitable frustrations of adult life? Is it true that the solution to violence lies in banning war games? Do war games make children (or those who like them) aggressive because they give vent to their natural aggression?

The hypothesis concerning drives offers an answer to all these questions: aggressive drive (like all the other drives) is innate in animals and comes to the surface *from within*; it doesn't necessarily come from outside. As a consequence, nothing can make it totally disappear. It can be controlled to a certain degree, like the sex drive, but never completely repressed. At least, not without paying a high price: the famous Austrian psychiatrist Sigmund Freud (1856–1939) already showed the disorders produced by repression of the drives that, from the subliminal level, struggle to emerge and become reality, and are restored, however, to the subconscious.

Konrad Lorenz garnered much criticism for giving credence to aggressive drive in our species, a drive that has been reiterated by a disciple of his, a specialist in human ethology to be exact. This is Irenäus Eibl-Eibesfeldt, and we next take a paragraph of his, enlightening for the clarity with which it extends the principles of ethology to the human being: "Many examples indicate that man is also subject, in several aspects of his behavior, to the manifestation of an excited state that is hard to control, since its origin is internal and subconscious, and that this materialization of an excited state, along with other motivating factors, decisively determine his inherent predilection to action. This is probably what happens with the instinct for aggression, which normally finds very few opportunities for release in present-day human society."

This very unpleasant implication of ethology has made many look upon it with antipathy, as if the hypothesis (or its defenders) were to blame for human aggression. Must science, in order to elude criticism, abandon this field of investigation only because its conclusions cause displeasure, and transform the biological roots of violence into a topic considered taboo (or leave it in the hands of those who speculate without any scientific basis at all)? Certainly not, since if science eschews fundamental questions, what then is its role?

Among the thinkers who in the course of history have maintained that man is by nature good is the Frenchman Jean-Jacques Rousseau (1712–1778). This philosopher reacted against the ideas of the French Enlightenment, which attributed all the evils of men to ignorance and advocated their salvation through science and progress. Rousseau, on the other hand, believed that *enlightened* education, instead of making men happier, corrupted them, and therefore demanded a return to nature; his heartfelt motto: the primitive or natural state of man is the ideal.

Freud thought differently after having witnessed World War I. Moreover, Freud was Jewish, and he knew by raw experience just how far human evil can go (he had to emigrate because of his Nazi compatriots and died in London). In his book, *Beyond the Pleasure Principle,* he put forth the idea of a new mechanism that prompted human behavior. The unconscious not only seeks out pleasure, as he had claimed up to that point, but in those recesses of the mind also dwells the *instinct of death,* which pursues the dissolution of life, the return to the lifeless. Even though it cannot be scientifically supported that matter has the will to abolish life, we can accept Freud's idea that there was something sinister in the nature of the human being.

There is a way to verify which of these two stances, Freud's and Lorenz's, or Rousseau's, is more exact. We can study peoples, who to this day and in complete harmony with nature, still practice hunting and farming, without owning land or having been *enlightened* and corrupted by education. Maybe we will find the *noble savage* among them, happy and peace-loving. We can also study humanity's past, in an effort to discover when violence first appeared, and if it already existed in the Prehistoric Age. Finally, we will take a look at the behavior of our closest relatives, apes, to see if violence is a part of their lives.

A topic of almost endless discussion among cultural anthropologists is whether violence is an occurrence in all known cultures, independent from their type of economy; by violence, we mean *intraspecific aggression,* meaning, among humans. It has been argued that some *primitive* societies have not known violence, like the Eskimo (or Inuit, in their own tongue) and the Bushmen (also known as San, even though this is a derogatory name used by their neighbors, the Nama). As we saw above, Konrad Lorenz and Irenäus Eibl-Eibesfeldt seem to come down on the side of the universality of aggressive drive as inherent in humans, or at least they believe that no one has shown that there is a single human group where no form of aggression will not manifest itself.

It is hard, nevertheless, to categorically avow this point of view, because aggression as a drive is one thing, and the ways in which it can manifest itself (ranging from challenging words to taking action) is another. The most direct form of aggression, of course, is to cause physical harm to another person, and the truth is that there are very few statistical studies on the few nonproductive economies of the world that have survived willy-nilly almost until modern times; nonproductive economies are the ones that do not produce food, but gather or hunt it instead. We have some data, nevertheless, on the Ache of Paraguay, the !Kung of Kalahari (the best-known group among the Bushmen), and the Yanomamo or Yanomami Indians, although the latter are, in addition to being hunters and foragers, cultivators of small patches of the jungle that they later abandon. The Bushmen, as previously mentioned, are considered a very peaceful group, while the Yanomamo, on the other hand, are considered highly aggressive.

In the Ache and Yanomamo societies, there are many known cases of murder, even though, surprisingly, the Yanomamo Indians, believed to be highly aggressive, have a homicide rate much lower than that of the Ache. In the case of the Bushmen, the statistics on violent death make no distinction between an accidental death and homicide, but the total for both categories shows an even lower number than that for the Yanomamo: when it comes to the question of homicides, the numbers will always be small. As for statistics, when the Ache lived in the forest, far from civilization, prior to anyone *making contact* with them in 1971, 70 percent of the deaths were attributable to violent causes: murder or accident; among the Yanomamo, the figure was 20 percent and just 11 percent among the !Kung.

At any rate, one has to keep in mind that many of the homicides among the Ache are infanticides, deaths of unweaned babies carried out by either of the two parents or by other adults (generally with the intention of accompanying an adult to his or her grave); maternal or paternal infanticide is a practice that is also known among the Yanomamo, although adequate statistics are not recorded by these Indians. Besides, more than half (60 percent) the homicides among the Ache was committed by their neighbors: farmers and cattle breeders. The Ache have lived throughout the twentieth century in constant armed conflict with their Paraguayan neighbors, who invaded their lands for the purpose of cultivating them; this is a situation that almost inevitably has arisen between hunting and foraging populations and their economically productive neighbors. In light of these two considerations, the figures for internecine homicides among the Ache and Yanomamo are much more closely aligned. All in all, the number of Ache who died at the hands of another member of the group was significant: 22 percent of all deaths were internecine murders, a figure that would seem terrifying even in the most marginal neighborhood of any large Western city.

With respect to the Bushmen, Eibl-Eibesfeldt, one of the authors who has studied their behavior most profoundly, reached the conclusion that they really are a peaceful society, even though violence is not unknown to them. Apparently, the Bushmen have the same aggressive drives as all other human beings, but they have learned to control them.

Prehistoric Cannibalism

Among human fossils we find examples of trauma, which could be the result of hunting or other types of accidents, but also aggression from other humans, including humans of other species. However, in the collection from the Sima de los Huesos site, at Sierra de Atapuerca, which contains the remains of about thirty individuals, only one case of a fractured bone is known, which then knit together and left a callus on the surface. Maybe the frequency of broken bones was not as high as had been assumed. We think that our friends from Sima de los Huesos formed part of a single community, and it is possible that they lived together and died at almost the same time (about 350,000 years ago). Therefore, the sample fossil from Sima is an exceptional one for studying extremely ancient populations. From among the human fossils from the Sima digs, there are many that show marks on the skull from small blows, but no fracture. In the case of one child, the blow was quite strong and depressed part of the frontal region up to the left eyebrow. Perhaps the poor boy was left with one eye, but he survived for a time because there are traces of bone regeneration. Unfortunately, in this case, as with all other human fossils, it is impossible to know if the thump on the head was delivered by another human. There is, however, one type of bone evidence that leaves no doubt about its origin: signs of cannibalism.

The paleontologist studies events that he has not witnessed, and therefore, he bases his conclusions on traces of evidence, much like a detective. In their diversity, paleontologists concern themselves with many aspects of life from the past, generally not at all connected with detective work. However, when the paleontologist is a specialist in human fossils and interested, moreover, in prehistoric cases of death and cannibalism, he wears two hats: part detective and part paleontologist.

Speaking of paleontologists and detectives, a saying comes to mind that is attributed to the French police, which says that you only see what you're looking at, and you only look at what you already know. Which in our particular case, translates to mean that at the moment of identifying cases of cannibalism at prehistoric sites, the thing to do first is to ask for what kind of clues, evidence of that practice, must you look. To do that, it is also useful to analyze what motives can drive human beings to eat one another. Thus, cannibalism could be divided, in broad swaths, into: (a) ritualistic cannibalism, when it is a funereal practice, symbolic in nature, which consists of respectfully consuming the bodies of deceased members of the group (in a kind of mystical *communion* more or less); (b) sustenance cannibalism, which consists of killing and eating members of other groups, merely for purpose of food (that is to say, as if it were a question of animals as prey); (c) cannibalism of members of the same group, dead or alive, under exceptional circumstances of dire necessity. With respect to evidence, it can be found in the fossils themselves or in their context (direct or circumstantial evidence in the language of detective stories).

On many occasions, the proof that is cited for presenting a case of cannibalism is based on the ways in which the human bones appear broken. But these fractures can be the result of multiple causes, of human or natural origin. For example, the crania of the famous *Homo erectus* of Zhoukoudian (near Peking, China) or Java, show the base of the cranium to be broken, which was interpreted as proof of the removal of the brain by other humans. However, today it is clear that the missing base of the cranium is a result of the fact that it is a fragile area, which is rarely preserved. Other causes of fractured bones can be the result of animal activity (trampling, scattering, eating) around the fossil prior to nature providing it a burial: the weight of sediment or falling rocks (in caves) on the bones, getting dragged along by rushing water, and many other nonbiological agents.

One type of immediate and conclusive clue would be to find cuts or lacerations on human remains like those found on herbivorous animals eaten by humans. To separate the meat from the bones, prehistoric man used his stone tools' sharpened edge, which he applied to certain spots, especially the tendons, leaving laceration marks on the bones. Very similar marks to these can also be produced naturally, and therefore the only convincing evidence comes from a group of such marks that displays systematic work with respect to specific muscular incisions. Not even then would cannibalism be proved, because the flaying could have had another purpose, perhaps a kind of ritualistic preparation or cleansing of the skeleton.

A wonderful example of the problems we run into in the search for ancient cannibals is the one offered by the digs at Krapina (Croatia). This is a site where probably some 130,000 years ago the remains of more than twenty Neanderthals piled up and were excavated between 1889 and 1905. To its discoverer, Dragutin Gorjanović-Kramberger (1856–1936), it seemed that the human bones were the remains from cannibal feasts, judging by the type of fractures that presented themselves, cut marks indicating slaughter, and burnt bones, being among some of the evidential aspects. Even though some modern authors are in agreement, others attribute the fractures to natural causes or to the methods of excavation in use at the time (contracting local help, and the use of such expeditionary techniques as picks, shovels, and explosives). Nevertheless, there are, it seems, genuine cut marks of human origin in Krapina, which do not necessarily imply cannibalism, as has already been noted (they can be the result of the ritual preparation of bones).

Other cases of flaying are found in the 600,000-year-old Bodo cranium (Ethiopia) and a more fragmentary cranial remnant from the Klasies River Mouth (South Africa) from between 60,000 and 85,000 years ago.

Particularly instructive is the case of the Engis 2 Neanderthal cranium, which many scientists believed was a victim of cannibals, but whose presumed cut marks were really produced in the laboratory by those preparing the fossil, and even some by the researchers with their anthropological instruments.

Tim White, from the University of California at Berkeley, is the author who has most rigorously studied the condition of bones after a cannibal feast and investigated skeletons from groups of pre-Colombian North American Indians who had been eaten by other Indians, and not necessarily as a matter of ritual. The same pattern is found at the French site in the Moula-Guercy cave dating back 100,000 years, where at least six Neanderthals (one was a boy of six and the others were adults) served as fodder for cannibals.

However, the characteristics observed by White are found in human fossils from the Gran Dolina site, at Sierra de Atapuerca, with an antiquity of around 800,000 years. Numerous remains show cut marks, patterns of dismemberment, and fractures similar to those of herbivores that had been devoured at that very same location. It doesn't give the appearance of a ritual practice, because the bones were abandoned after the feast along with the osseous remains of normal prey. Instead, it looks like a case of cannibalistic feeding, and it is quite possible that anthropophagens (cannibals) did not feed off abandoned human cadavers, nor their dead pals done in by other causes, unless they had killed them for the purpose of eating them. Among the victims, six at least, are two children around four years old, another approximately ten, and another around fourteen.

All this seems supremely negative and gives the impression that we humans are really profoundly terrible beings, more than any other animal species. Are we humans, in fact, so violent? Let's start by making a distinction between intraspecific violence (within the same species) and interspecific violence (among species). There is the false impression that vegetarian animals, that don't kill animals of other species for the purpose of eating them, are peaceful toward their own, and from this, some deduce that vegetarian humans will behave likewise. Well, it isn't so. Vegetarian animals, beginning with the dove, the symbol of peace, can be greatly intolerant with their own, to the point of causing their death, while we do not normally detect much aggression in a group of lions peacefully basking in the sun.

Not to go too far astray in the animal kingdom, let's observe our closest relatives: the other monkeys. Some, like gibbons, form minimal social units constituted of a single male and female, and live together in perfect harmony and exemplary mutual happiness alongside their young; of course, the pairs are territorial and very aggressively defend their vital space against other pairs. Other species, as in the case of gorillas, live in groups formed by a single male and several females. Here, harmony also reigns, even though the male violently defends his space against an encroaching rival. Neither of those scenarios applies in our case, because we humans live in groups formed by many males and many females.

In this type of group, rigid hierarchies are often established among the males, who spend the entire day threatening and attacking one another. Many visitors stop by the baboon den at the Madrid Zoo, because there are constant skirmishes between the males and also between the females (who have their own hierarchy), besides the aggression the males display toward the females. Among chimpanzees, aggressive behavior is not as frequent, even though there is some of that too.

If we compare ourselves to the baboons in the zoo, our behavior is extremely amicable. Men and women pass each other on the street everyday and work side by side at their jobs, and violence, or even a threatening gesture, is quite rare. In fact, if it were necessary to define *Homo sapiens* socially, we would say that it is an extraordinary species that forms very broad groups of males and females who live together with barely any physical contact, and indeed individuals who did not know one other before often come into contact with each other. Moreover, we pair off in order to procreate and look after our young; ours is a two-tiered society: the family and the group.

When Konrad Lorenz and other ethologists began the scientific study of animal behavior, they observed that at the core of the groups, aggression had limits placed on it: the weakest individuals and the vanquished emit signals of appeasement during combat that prevent the strongest (the victors) from ending their lives. It is often sufficient to adopt a fetal pose to stop the aggression, because the young cannot be attacked. Lorenz and his colleagues asked themselves what happened in human evolution for aggression to have become a recourse and death the frequent result. Could it be that we lost our rituals of appeasement? Recent studies have shown two things that contradict the observations of the ethologists; in the first place, witnessing acts of harmful aggression in animals is just a matter of a few hours of observation, before or after they manifest themselves; secondly, when accurate statistics of aggravated assault and brutal crimes are recorded, the numerical proportion of citizens is lower in the worst neighborhoods of New York than in wolf packs or flocks of doves.

In this line of reasoning, there is, however, a trap, which perhaps the reader has noticed. The statistics for citizens deaths are low during times of peace, but they dramatically shoot up if we include wars. All social and territorial animals defend their homes from other groups, whether the encroaching species be carnivorous or vegetarian. A lot is at stake because territory means resources. Male chimpanzees, which innocently feed off fruits and leaves, turn ferocious to the point of exterminating the males of neighboring communities.

But in conflicts between animals the rule of blood, or genes, is followed, and relatives band together much more the closer they are related to each other. In this respect, our species is clearly a special case, because during great ideological confrontations, those who do not know each other will come together to kill their earthly brothers. We are a species that creates symbols, forms groups that base their identity on shared symbols more than on shared genes, and is emotionally linked to those symbols. This is how evolution fashioned us, and the result is that we are too easily manipulated: the one who controls the symbols also steers our emotions.

The Apprenticeship

The empirical Scottish philosopher David Hume (1711–1776) found that there was an unsalvageable distance between common sense and logic. The first allows us to establish links between two events when one immediately succeeds the other. *We know* that the sun will rise again in the morning, and that the stone that we toss into the air will fall back down to the ground. It is predictable that one billiard ball striking another sends it into motion at the moment of impact. Better yet, in addition to linking events, on some occasions we are capable of establishing causal relationships, whenever two events occur in the same location and close together in time, and one (the cause) immediately precedes the other (the consequence).

But, no matter how much those sequences of events are repeated, and no matter how tired we are of seeing them, there is no logical reason, thought Hume, for being certain that day will necessarily follow night, that the stone will inevitably fall, or that the billiard ball that is struck will be set in motion. We tend, out of custom or habit, to establish relationships between incidents that occur successively in the same place. It is not a question of *truisms,* like mathematical deductions that are drawn from axioms, but rather *cold, hard facts*. It isn't impossible, in pure logic, for the sun not to come up in the morning; on the contrary, two plus two is necessarily four. The consequence is that what we say we know is only our belief, because it is not the outgrowth of logic, but of intuition based on the repetitions we observe with our senses. Mathematical truths are irrefutable, but unfortunately they have to do with an abstract world that is not reality. This dependence on the senses, and not reason, in getting to know the world that surrounds us, is highly critical, because the senses are the indispensable door to understanding. For British empiricists like John Locke (1632–1704) and David Hume, the mind of a newborn is like a blank sheet of paper or tabula rasa, whereby there is nothing in the mind that has not been previously perceived by the senses (*"Nihil est in intellectu quod non antefuerat in sensu"*). In this way, just as the most radical idealism ends up denying the possibility of attaining a true knowledge of reality, with Hume, empiricism also ended up falling into the deep abyss of skepticism.

Konrad Lorenz rebels before Hume's skepticism, saying that, thanks to evolution, organisms live in the world with a true knowledge of what surrounds them, for their ancestors would not have survived otherwise. We are not born a tabula rasa. That knowledge is recorded in the genes. François Jacob, as we saw, referred to the genome as a system of memory, and Lorenz considered it an apparatus for the acquisition of knowledge, a cognitive apparatus. The perfect adjustment between organisms and their medium (which includes all the other organisms too), that is to say, their perfect adaptation, provides an exact consonance between the eye and light, the ear's tympanum and sound waves, the wing and air, the fin and water, the butterfly's quivering proboscis and the calyx of the flowers that it sips.

However, that innate knowledge, anterior to all experience, cannot foresee all the circumstances of an animal's life. It is very appropriate for the animal to accrue experience in order not to repeat its mistakes, but the genome cannot be changed once a new being has been conceived. The genome's method for gathering knowledge is not based on modifying itself during its lifetime, since, on the contrary, the information acquired is the result of natural selection: the organisms that perform well survive, those that do not die, in the same way that those equipped with beneficial structures are successful, and those that swim, fly, or digest poorly produce fewer offspring. Unlike the individual, who learns as much from his mistakes as from his successes, the genome only learns from its *successes*.

But there is one thing that genes really can do for the individual: they can equip him with the temporal ability to gather knowledge in his nervous system. That knowledge will not be transmitted in the genes, but it will help the individual to transmit his genes. In man, knowledge travels from one generation to the next via culture, and thus, it doesn't die with the individual; but also in some animals, there is a certain transmission of information via this extragenetic route. Chimpanzees, to cite our closest relatives, have territorial traditions, which pass from parents to children, in the use of instruments, and in the case of some species of songbirds, regional dialects are recognized in the melody.

But these animal cultures are differentiated from human cultures in that they are not transmitted via language, that is to say, communication by symbols, and therefore they are incomparably poorer. As Lorenz says, while in the human being, information that is inherited through culture is much greater than that which reaches us via the genetic route, in all animals, without exception, exactly the opposite happens: there are many more *bits* of information in inherited genes than in behaviors learned from the parents.

In animals, the instructions for acquiring knowledge are supremely rigid on some occasions. When a worm comes into the world, it is programmed to recognize as its mother any large object that is moving in its vicinity, whether it be a person or ball. This type of programming is simple and highly useful, and eschews complex descriptions. Likewise, we humans would also equip a group of hypothetical self-replicating automatons with a set of basic instructions were we to launch them into outer space for the purpose of conquering it. If the object that is normally found near a worm when it comes out of its cocoon is its mother, what need is there for genetically inscribing into the newborn a complex pattern of recognition from an adult female worm?

It was Konrad who discovered these innate processes of learning that give the appearance of strong impressions engraved during infancy, and he called them *imprints*. Sometimes the imprint of an object is produced, which is what happens to the newborn worm. That printing takes place during a particularly susceptible stage and, moreover, is irreversible. A male duck raised with chickens, worms, or ducks of various species, and then set free, will prefer a sexual partner from the species with which he was raised before choosing one of his own. It is interesting to see just how strong the imprint is: it only acts as an emancipator of a fixed reaction; a female rook raised by Lorenz preferred him for a father or sexual partner, but not as a social companion, and in order to fly, bonded with other rooks of her species.

Besides being an object, the imprint can be a motor pattern, that is to say, a sequence of movements like those that produce a bird's melodies. Even if you were to isolate year-old finches in September, in other words, long before they begin to sing, what they have already heard as chicks suffices for them to reproduce the warbling of the species the following spring without any problem. The warbling imprint is permanent.

Even though the above examples of imprints come from species of birds, where they are easily recognized and quite obvious, learning by means of imprint also shows up in mammals, including primates and possibly the human being as well. Freud already understood the possible consequences in the early and most sensitive stages of life from all that surrounds the child, beginning with his or her parents.

The imprint does not necessarily determine who can be one's sexual partner the day after tomorrow, but also who it cannot be. Edward O. Wilson, the sociobiologist, finds the most conclusive proof that instincts also exist in humans precisely in the taboo of incest, even though not necessarily in the form of rigid programs of behavior that do not permit modification, but rather as innate predispositions. The aversion to incest, which would be a natural feeling, probably was reinforced culturally through laws that prohibited it and elevated its avoidance to the category of a moral norm.

What natural explanation is there for the rejection of incest? That of avoiding procreation of offspring with serious problems resulting from consanguinity. According to some estimates, we humans are carriers, in general, of one or more lethal recessive genes at any locus of our twenty-three pairs of chromosomes. The fact that we are alive is explained because in our genome we have two pairs of genes, inherited from each of our parents. The lethal recessive genes only produce death when two are joined together, one via each parent. If we only inherit one (from the mother or father, it makes no difference), the lethal recessive gene remains hidden. However, since we share half of our genes, on average, with each one of our siblings, there is a real risk that if we procreate with any one of them, two lethal genes could come together in one of our children, and the child not survive. That is a one eighth probability, if we are talking about one lethal gene in only one locus of the twenty-three pairs of chromosomes, but the probability increases to one fourth if we are carrying two lethal genes (in two different loci). The same probability holds true in father/daughter and mother/son relationships, where half of the genes are shared, and therefore, the taboo of incest also extends to those situations that are not considered morally acceptable in the majority of cultures. Of course, there are exceptions, as for example with the ancient Egyptians, where marriage between brother and sister was prescribed, not prohibited, for the sake of preserving the purity of blood in the pharaohs' family (but it also took place among the general populace); the fact is that given the enormous diversity of past and present human cultures, it is difficult, perhaps impossible, to find any universal characteristic.

How might abstention from incest have evolved and become a dominant taboo? In instinctive behavior, as in any other trait with a genetic basis, there are individual variations. Those members of the community who were less given to practicing consanguineous sexual relations, probably found themselves less penalized with their offspring, and thus the instinct to shun incest presumably spread to the entire population over time.

The instinctive aversion to incest has been verified in situations where children born to different parents live together as brother and sister from a very early age. A man and woman generally do not feel sexually attracted to each other if either of the two was younger than thirteen when they met. It is as if we are born with the instruction (whose imprint is forever stamped into us) not to procreate with anyone with whom we have lived since childhood; that feature probably can never be erased. One well-documented case study, in this regard, is that of the Israeli kibbutz, in which boys and girls attend school together, but where marriage between former classmates does not occur despite any lack of opposition from the parents (who probably see no reason to oppose it). There are similar examples in other kinds of cultures.

Strangely, Freud held a diametrically opposite point of view, since he based the cornerstone of his theory on the so-called *Oedipus complex,* which claims that the son falls in love with his mother in infancy and looks upon his father as a rival from that point forward. The reason for the existence of laws that prohibit incest is, argued Freud, precisely because it is a latent desire against which society deems it necessary to protect itself. Is incest something toward which we feel instinctively inclined or something we find absolutely abhorrent? No two hypotheses can be more incompatible with each other. We could add another one that says incest is prohibited because we humans know, rationally and instinctively, that it is not seemly (even though it is true that in many cultures that forbid incest, there seems to be no awareness of its pernicious consequences).

However, animals, including monkeys, also eschew incest in communities where it is not common, under normal conditions, for mothers to copulate with their sons or for siblings to copulate with each other. Moreover, in social mammals, sons or daughters frequently leave the fold upon reaching sexual maturity (the females do it in the case of chimpanzees). All this makes one think that aversion to incest is a response to a genetic predisposition to eschew sexual contact with persons with whom we have lived since childhood . . . and therefore, in this case, Freud was mistaken.

But animals also have the ability to make associations between occurrences, that is to say, to understand the norms of nature. Human logic is merely, after all, another way (analogous) for creating causal links between incidents. We can trust, despite what Hume says, that ability animals have for ascertaining associations between occurrences, because it was natural selection that was responsible for its existence.

One year at the Atapuerca digs, we had a young fox that used to creep closer to get a better look at us (youngsters are always more curious, more eager to explore, than adults). We began to toss him bits of food when it was lunchtime, and we immediately got him into the habit of visiting us at exactly that time of the day. Several weeks later, after having completed our work, it was purely by chance that we went back to the site precisely at lunchtime, and there was the innocent little fox, waiting for us. It is obvious that the animal *knew* he could get food at that time of the day and at that spot, just as he *knew* how to explore other sources for food (hunting, carrion, garbage, etc.) at different times and in different locations. But how, or rather, by what mechanism, had he learned this?

In a famous experiment, the Russian physiologist Ivan Pavlov (1849–1936) discovered that animals establish associations among stimuli. Setting out food for a hungry dog (or a person) causes him to salivate and stimulates the action of the gastric juices. Pavlov rang a bell whenever he set out food for the ingenuous laboratory dog. In this way, he achieved the physiological reaction, which prepares the animal to come for the food, to be produced with just the bell's sound, without there being any need at all for the presence of food. This association between two stimuli (the food for one, the bell for the other) is known as conditioned reflex.

Pet owners (and to a greater degree, professional animal trainers) know very well that a specific behavior can be conditioned by the use of a reward or, inversely, suppressed using the threat of punishment as the basis. This type of mechanism, called *instrumental conditioning,* would be for the American psychologist Burrhus Frederic Skinner (1904–1990) the key to understanding animal behavior as well as behavior in humans. According to Skinner and the school of *behavioral science* that he headed, animals, in fact, probably neither *knew* nor *understood* anything. Simply, the frequency with which a pattern of conduct is repeated increases with the reward, and conversely, the probability that the pattern will manifest itself diminishes with punishment.

Modern *cognitive psychology*, which has replaced Skinner's behavioral psychology, believes, on the other hand, that animals are formed, in some unknown way, by internal representations of events. These events can be either stimuli that they receive (reward, reprimand, or a neutral stimulus) or their own motor patterns or responses. When animals perceive that there is in the external world a relationship of dependence between two stimuli, or between a stimulus and response (or, vice versa, between a behavior and stimulus), they somehow form associations between the internal representations of those events. In other words, they *know* that a specific activity will produce either a reward or reprimand, and that a specific stimulus means either something good or bad.

We now enter a much more complex terrain than that of imprints, and it is very difficult for us to imagine how we could equip a self-replicating machine with something similar to those internal representations, such as the possibility of having them establish associations among them. However, such abilities are found in animals that have a really simple nervous system. There are even more surprising things that they know how to do.

Chapter IX

The Marvelous Adventures of Baron von Münchausen

Effulgence

Konrad Lorenz observed that animals possess a series of astonishing faculties, each one of which is present in man, but which are integrated in us in order to produce the rational mind. Those abilities that animals demonstrate, to a greater or lesser degree, constituted a necessary condition for human thought to surface at some moment in the history of life, like an effulgent beacon of light in the evolutionary process. They were the roots of abstract thought. Even though the faculties in question may undeniably be explained one by one, they do not form a historic sequence, since they did not successively appear, but rather materialized independently of one another and functioned separately, until they converged in the first thinkers. From their association emerged something much more important than their mere totality; new properties appeared that did not exist at the inferior level: a conscious, symbol-oriented mind and language. Here we find those animal faculties upon which human thought established itself.

Animals are, surprisingly, capable of distinguishing between what is essential and what is nonessential in the objects that form their world. We are struck by the fact that animals distinguish friend from foe, or identify their food, even though those objects can present themselves in many different forms. Let's imagine that we want to create a mechanism for a system to recognize a type of animal, a dog, for example, in the multiple stances and different settings that can pop up, and under the enormous spectrum of recognition that is possible. It is really difficult, in practice, to achieve such prowess, because it is a question of nothing less than equipping the system with the ability to recognize the essential canine characteristics that are always found in each and every dog. However, this is an ability to deduce, to objectify, to create categories (even if not consciously) that possess a multitude of animals. We also find this ability in the youngest children who, when they see a dog, of any breed (mastiff, greyhound, or poodle), say "bow-wow."

Likewise, many animals have a very uncanny ability to orient themselves, when out in the open, thanks to their vision. This faculty is more developed in species where open space presents a broader spectrum of complexity, for example, in arboreal mammals. Predominate among them are primates, which must execute accurate movements in grabbing onto tree limbs when moving from branch to branch. In New and Old World monkeys (*catarrhines* and *platyrrhines*), the eyes are set in the anterior position, thus affording the animal an ample field of three-dimensional (or *stereoscopic*) vision. This binocular vision (with the two eyes) inherited from our tree-dwelling ancestors also makes it possible for us humans to have broad spatial perception.

It is hard to admit that abstract thought could have developed in a species that did not perceive the world in images, but rather, like the majority of mammals, in smells. We have proof of this in our use of language, in the number of times that, in expressing any idea, we use adverbs and other words to indicate spatial position or movement (such as when, for example, it is said that abstract thought is a *superior* ability that we acquired when we *ascended* the scale in evolution, an ability not possessed by the ancestors we left *behind*).

A family's pet cat has climbed on top of a wardrobe. From there she takes in the rest of the room, shifting her gaze from one part of the room to another; the window, table, chair, dresser, bed, floor, and so on, are the objects of her investigation. She seems to be thinking about her next move, calculating the distance that stands between her and several possible landings. At last, she fixes her gaze on one particular spot: she has already made up her mind; seeing her muscles contract like a spring coil and her vision not stray a single millimeter from her objective, we can predict where she'll land.

When some types of animals study their surroundings before executing any action, it could be said that that they are feeling the surrounding space with their vision, gathering sufficient information before acting. Konrad Lorenz imagines, in the case of mammals, that an "internal action" is probably carried out in the "imagined space," meaning the space represented as a model inside the central nervous system. All human thought, such as it is, can only be conceived as an "action in imagined space" whose development takes place in the central nervous system. Before leaping into action, the animal equipped with such abilities (as also happens in man), carries out an internal simulation of what would happen under the different options that present themselves. In other words, the animal carries out mental experiments. It imagines possible outcomes and chooses one of them.

For François Jacob, one of the keys to understanding the human being lies precisely in *imagined space*. The faculty that enables reality to be represented in a visual form allows "the memorized images of past events to be separated into their different components, to be brought together again at a later time and to produce new situations and previously unfamiliar representations; herein lies the ability not only to preserve images of past events but to imagine new possibilities and, therefore, to invent the future."

It is inferred from all of this that learning and memory constitute an indispensable basis for abstract thought, which is how Konrad Lorenz sees it. Those *memorized images*, spoken of by François Jacob, have to be stored somewhere so they can be pieced together later and thus create future plans. In the delightful *The Jungle Book* by Rudyard Kipling (1865–1936), the monkeys (the *Bandarlogs*) are the object of the *wise* animals' scorn, like Baloo, the bear, or Bagheera, the black panther. They attempt to be like humans, and form societies with leaders, but they fail because they are unable to retain a single idea in their heads. Many projects begin with great expectations, but any trifling matter, like a coconut falling, distracts them and makes them forget their project. It is imperative that the head must not be *hollow*, since it must preserve information, at least for a certain period of time.

In a medium filled with obstacles, the exploratory ability of binocular vision would be worthless if it were not accompanied by an enormous motor flexibility with a great variety of movements. On the other hand, animals that live in barren environments have no need for such vision. On the immense, open sea, the captain of a ship only needs to adjust the speed (to slow down or speed up) and steer the vessel (to navigate his course). At the other extreme, canoes descending the Sella River require a great deal of maneuverability in order to make their way around obstacles and whirlpools.

A horse, an animal adapted to the open plains, can move at a slow pace, trot, or gallop, but needs a great deal of patience to execute the simple maneuvers in classic equestrian competitions. Even the marvelous exhibitions given by the equestrian schools of Vienna or Jerez are nothing compared to the versatility of the monkey. Let's return again to this primate.

Curiosity, inquisitive behavior, is another one of the conditions for abstract thinking. No one knows what produces that appetite for information, which implies a desire for exploration, but it is clear that it is only possible when the animal is in a *relaxed* state, that is to say, when he is not subject to a more *pressing* anxiety in connection with the search for food, flight, reproduction, care for the young, etc. Under those circumstances, the animal is not in a *light-hearted mood*, but when the urgency of the moment passes, it is then possible for it to search for new things and also find some time for playfulness. Mammals, in general, and monkeys, of course, are very *inquisitive*, especially during their earliest years, and man remains so his entire life. In that sense, we are forever child-like. The tendency to investigate when circumstances permit, that is to say, when there is no greater or more urgent preoccupation before us, involves an enormously effective way of gathering information about the world that surrounds us. That database remains latent, in a dormant state, and is only called upon in critical moments. The information that an animal stores up, about the mysteries of its territory during its aimless wanderings, can save its life when at a specific moment and without time to *think*, it is being pursued by a predator and has to flee.

The game, if properly viewed, is the opposite of operative conditioning, wherein a simple movement (a basic hereditary coordination) is put to the test against different objects, until reinforcement is produced either by way of a reward or an aversion to punishment. The rat in Skinner's cage can, just by scratching, accidentally depress a lever and immediately either be rewarded with food or punished with an electric shock. That action of depressing the lever will from that point on either be repeated frequently or abandoned. The rat can also learn to find the reward that is hidden at the end of a labyrinth simply by moving through it. But in the game, many kinds of complex hereditary movements are employed in rapid succession and toward the same object, like when a cat first imitates the behavior of the hunt, then that of engaging in a struggle with another cat, and later, that of defending itself against a dog, etc. The game is, therefore, somewhat different from operative conditioning, and it involves very important *training* (and therefore, younger individuals especially practice it) for the *serious* situations of adult life. In our case, we never completely abandon the desire for play.

But in monkeys, always in monkeys, exploration is not limited to the external world, but includes their own bodies, and is transformed into self-exploration. How can that be? Well, because the primate has his hand located within his field of vision when he grabs onto a branch or handles an object. Both Konrad Lorenz and François Jacob equally reason that seeing the prehensile hand next to an object, and receiving sensations from the hand as well as the images that appear along with those sensations, a fruit, for example, whose touch is received by the hand, is what allowed one of our ancestors to identify his own body as one more object in the real world. When the monkey seizes the hand of a companion between his own, he understands to what point they are equals, and in this way, he sees himself reflected in the other and discovers himself as if he were standing in front of a mirror. This is how our ancestors glimpsed their own existence, their reality as one more object in their environment.

The discovery of the objective nature of the body was another premise for abstract thought, and apparently chimpanzees who, pursuant to certain experiments, recognize themselves before a mirror, also support that premise. The awareness of oneself, considered as an object, affords the human being the possibility of imagining himself inside his own *imagined space,* seeing himself *from the outside.* It even opens up to him the terrifying possibility of *not seeing himself,* of imagining a time anterior or posterior to our existence.

Another necessary basis for the appearance of language was probably the capacity for imitation, which once again is so highly developed in monkeys, but which is also found in some birds that learn their singing by imitation (totally, or from an innate model).

Lastly, Konrad Lorenz also points out, as a prerequisite for the effulgence from which abstract thought is born, the existence of tradition, or in other words, the possibility of the transference of information from one generation to another. Examples of regional dialects occur among birds that have to learn to warble, and again among primates, and in particular with Japanese macaques and chimpanzees, we are familiar with impressive cases of culture, learned in the broad sense of transmission of habits through a method different from that of the genes (in the strict sense of perpetuation of ideas, culture exists among us only because it is needed for symbolic communication, in other words, language).

Among chimpanzees in the wild, thirty-nine patterns of behavior are known (which have to do with the use of tools, hygiene, and courtship) that are normal in some communities and are completely absent in others of the same species. Those different habits cannot be explained on the basis of ecological circumstances endemic to each community, but are simply the result of a new pattern of behavior surfacing from time to time, and the members of the community practicing it and transmitting the habit from parents to children.

Darwin Already Said That

On the question of evolution, it is common to hit upon ideas (more or less sophisticated) that had previously occurred to

Darwin. Then, someone who has recently read him realizes it and reminds us of it: Darwin already said that!

Since Darwin, in 1871, was concerned with the origin of man in a book (bearing that very title), it turns out that some of Lorenz's ideas can also be found there, specifically in the chapter titled "Comparison Between the Mental Faculties of Man and Inferior Animals."

Darwin points out the commonality between man and animals with respect to certain instincts, "self-preservation, sexual drive, maternal affection for the newborn, the infant's propensity to be suckled, and other similar things," but also in the area of primary emotions, like pleasure and pain, happiness and misery, and even more complex ones, which Darwin considers rudimentary feelings in animals, such as jealousy, shame, self-love, vanity, and playfulness or the desire to play (on whose existence, as human as those feelings may seem, many pet owners would be in full agreement).

Next, Darwin takes up consideration of "more intellectual emotions and faculties, which form the basis for the development of superior mental aptitudes." As one can see, this is an idea very similar to the one expounded upon by Konrad Lorenz. Among these superior mental aptitudes are curiosity and the penchant for exploration—characteristic of many mammals—enthusiasm for imitation, focus and the capacity for concentration (and as an example, take the cat that is about to pounce on its prey), memory, imagination, and rationalism. With respect to imagination, Darwin writes in terms very similar to those of François Jacob, which we saw a little while ago: "The imagination is one of the highest prerogatives of man. By this faculty he unites former images and ideas independently of the will, and thus creates brilliant and novel results. [. . .] The value of the products of our imagination depends, of course, on the number, accuracy, and clearness of our impressions that we have, our judgement and taste in selecting or rejecting the involuntary combinations, and to a certain extent on our power of voluntarily combining them" (*The Descent of Man*, 1871, chapter III).

Darwin speculates on the highest mental faculties in man: abstraction, general ideas, self-awareness, intellectual individuality. He thinks that it is very difficult to know to what degree those characteristics exist in animals, but he argues in a way similar to Lorenz concerning abstraction in animals: "If one may judge from various articles which have been published lately, the greatest stress seems to be laid on the supposed entire absence in animals of the power of abstraction, or of forming general concepts. But when a dog sees another dog at a distance, it is often clear that he perceives that it is a dog in the abstract; for when he gets nearer his whole manner suddenly changes, if the other dog be a friend. A recent writer remarks, that in all such cases it is a pure assumption to assert that the mental act is not essentially of the same nature in the animal as in man" (*The Descent of Man*, chapter III).

With respect to reason, which he indisputably considers the primary faculty among all the mental capacities, it probably had to do with the ability to make associations between events, which he finds highly developed in mammals, but which also exists in other *inferior* vertebrates.

Of Keys and Locks

In a completely independent manner, the philosopher Daniel Dennett has expanded upon some ideas on the origin of our mind that have many points in common with Konrad Lorenz's thoughts (and with Darwin's). Dennett's model is a tower of ascendant complexity, in which some types of minds are floors built over other inferior, less efficient ones.

The bottom floor is occupied by what Dennett calls *Darwinian creatures*, which display the simplest forms of behavior and which were the first. These Darwinian creatures exhibit disparate kinds of behavior faced with a given problem; it is as if they were different *keys* trying to open the same *lock*. Several types of *keys* appeared through the blind mechanism of mutation and recombination of genes. Natural selection favored one of the variants, and it multiplied. Darwinian creatures only have one kind of response (each one a key) to a problem (the lock): some get it right and others do not, but there is no flexibility in the type of behavior, which cannot be modified in the least during the individual's life; the key with which one is born remains immutable.

In time, creatures appeared with a more flexible form of behavior (a *phenotype*, to be exact). It could be said that those creatures were unfinished products at birth, and thus were susceptible to modification by their circumstances in life. Presented with the same problem (the lock), these unfinished products would blindly attempt different kinds of responses (different keys) until one of them worked (opened the lock).

These organisms do not presume to be a great advancement on the tower of ascendant complexity, but some of them (as always, by pure chance, that is to say, by simple mutation and recombination) also made use of a mechanism that reinforced successful behaviors (the keys that opened the locks) and, on the contrary, did not allow those behaviors that led to failure (to a more or less complete disaster) barely, if ever, repeat themselves. In other words, the next time one of those creatures finds itself confronted with the same lock, in all likelihood it will first try the key with which it was successful on the previous occasion.

Those organisms learn from their experiences, and Dennett calls them *Skinnerian creatures*, because their behavior is the result of the mechanism of reinforcement or displeasure through reward or punishment, which according to Skinner forms the basis for the entire process of learning: it is what he called *operative* or *instrumental conditioning*. This form of learning, as you will remember, is different from how genes do it; the latter only learn from their successes (Darwinian creatures that possess the keys that open locks survive, the others die), while Skinnerian creatures also learn from their failures: the keys that do not open locks are eliminated without the organism that produces them dying in the process.

At times we humans are *Darwinian creatures*, and we can also be *Skinnerian creatures*, but we have a third system at our disposal for discovering successful behavior. We are *Popperian creatures*; this is the name of the third floor of Dennett's tower, which pays homage to the famous philosopher Karl Popper, for whom our design "allows our hypotheses to die rather than us." In other words, it is remarkable (almost magical) that multiple attempts at a lock with different kinds of keys are not carried out in real space, where every failure means suffering another blow and perhaps dying. In the event of some bad luck, a Skinnerian creature can take a long time in coming up with the correct key, for it will try all of them at random. The next time around, everything will be easier, thanks to operative conditioning, but the animal can come out of it so badly damaged from the experience that it is possible there will be no next time.

Instead of experimenting in external space, man and also other animal species have the capacity for experimenting in imaginary space, that is an internal space where the different keys can be tested risk-free. The first attempt by a Skinnerian creature faced with problem (a lock) is a random option, chosen by chance, among the different possibilities (keys) at its disposal. On the other hand, the Popperian creature tries, in the first place, the option that proved successful in his cerebral simulation and which, therefore, had been preselected. To do that, a certain capacity is needed, and with sufficient accuracy, for internally reproducing the external medium and for creating, in this way, a sort of internal environment.

In other words, contrary to philosophers who argued that we came into the world absent any knowledge at all of the real world (like a blank slate), and even that we do not possess the ability to acquire it, the fact is that Popperian creatures are born with an internal model, quite accurate for the awareness that it brings to them, of what there is beyond their brains. Moreover, they have the possibility of acquiring knowledge and gradually improving and completing in life that internal representation of the external world with which they are born. It goes without saying that the internal world is not a dollhouse, a scale model of the real world. As Dennett says, it is not a question of imagining that we have inside our heads a hot, imaginary oven that burns an imaginary finger that we stick inside. But all the same, there is sufficient information in that internal image to fulfill its function in keeping us from burning a real finger in a real oven in the real world.

According to Dennett, only invertebrates, and maybe not all, could be purely Skinnerian creatures. Vertebrates are, without exception, Popperian creatures.

The internal image of the external world can be improved by adding the experiences that are accumulated during the individual's life, and if possible, the experiences accumulated by others. The way in which some animals appropriate knowledge from others is through imitation, and tradition exists because of it. Culture, understood in its broadest sense as a nongenetic form of the transferal of information from one generation to another, follows Lamarck's mechanism rather than Darwin's; that is to say, it is culture that transmits acquired knowledge, not genes.

Through culture, we inherit the tools made by people who died long ago, and animals, like chimpanzees, do the same. In the cerebral simulations of resolving problems, tools can also be included, so that a real key can be tested with a lock inside our heads. In this way, not only the world of natural objects is incorporated into the internal model of reality, but also the world of fabricated objects, meaning *designed space*.

Daniel Dennett calls those creatures that have tradition *Gregorian creatures,* this time in honor of the British psychologist Richard Gregory. This author saw that a pair of scissors is not only the result of intelligence, but provides additional intelligence to the person who uses them. Tools are *potential intelligence*. The more intelligence invested in the tool's design, the more *potential intelligence* is conferred on the one who uses it. Groups of chimpanzees that do not know how to prepare twigs for *fishing out* termites (because it is not in their tradition) miss out on this important source of proteins. On the contrary, hominids that learned to break a bone with a stone gained access to the rich marrow it contains. The difference between those who use firearms and those who use wood lances is much greater. The rifle has more design, more incorporated intelligence, than a stick with a sharp point.

Obviously, only a Gregorian creature could transform itself into a human, but that is not to say it is easy to know how it accomplished that feat. Dennett thinks that Gregory himself holds the key: *mental tools* are among the most important tools, meaning words. The outcome of incorporating *mental tools* into our *imagined space,* and *playing* with them and combining them in an infinite number of different ways was, over time, the human mind. Also thanks to this powerful tool, this time serving communication, it was possible to learn not only from our own mistakes, but also from the mistakes of others.

Emergence

Baron von Münchausen (Karl Friedrich Hieronymus Baron von Münchausen) was a historic personage who lived between 1720 and 1797 and fought in the army of the Czar against the Turks, attaining the rank of captain in the cavalry. Münchausen won great fame in his lifetime as a teller of incredible tales, which were finally published in several books. In the one by Gottfried August Bürger, published in 1788 (*The Adventures of Baron von Münchausen*), the following tale is told. Once the baron wanted to leap across a marsh in a single bound on horseback, but he discovered halfway across that he had calculated badly, and so, being already in midair, he turned back to the original spot to attempt another leap. On his second try, he again fell short and found himself up to his neck in the marsh, not very far from the opposite shore. Fortunately, his strength and ingenuity saved him from disaster once more. "Undoubtedly, I should have died had it not been for the strength of my arm pulling me out by my pigtail, along with my horse, which I kept firmly under control by pressing my knees tight against his flanks."

The two great problems with the history of life are the appearance of the first living being and the birth of the rational mind. Common sense says that you cannot get something out of nothing, meaning, where there is nothing but emptiness, there is nothing that can emerge from it (and elms do not yield pears). Such a thing seems logically impossible. Someone or something from outside had to create life and intelligence, because it would be impossible for inanimate matter to organize or assemble itself in order to produce life. How could letters assemble themselves to write *Don Quixote*, or notes arrange themselves to write a symphony? In other words, only an Intelligence could produce *intelligence*. The contrary would be to think it possible to ascend into air like the imaginative Münchausen: pulling oneself up by the hair of one's own head; it would be tantamount to advancing the amazing (but incredible and surreal) adventures of Baron von Münchausen to the status of a scientific paradigm.

It is not necessary, however, to resort to a supernatural agent to scientifically explain the origin of life and consciousness, because it is quite possible to conceive of both properties as separately emerging, in one instance, from organic molecules and, in another, from noncognizant animals. The philosopher Daniel Dennett explains it this way: there are people who prefer to believe that *celestial skyhooks* exist that allow organisms to go higher on the tower of ascendant complexity, to which we previously referred. Each leap of ascension, from the origin of life to the human mind, would require the intervention of one of these skyhooks. But there is an alternative device for ascending: cranes. A crane is not a magical solution comparable to Baron von Münchausen's use of his long plait of hair. We use cranes every day in construction, but not skyhooks.

By crane, Dennett understands it to mean the appearance of a new property (through the ordinary Darwinian mechanism of natural selection) that, even though it may not have manifested itself to be used as a crane, nevertheless speeds up evolution and conquers new design spaces. Language, for example, would be a crane that would have accelerated the evolution of the human mind. Culture is a crane for building other cranes. In contemporary society, we are witnessing the appearance of cranes (like the silicon *chip*) that, almost by magic, accelerate *technological evolution* and create new, heretofore unbelievable machines. A long time ago, sexual reproduction was also a biological crane. It did not appear, of course, for the purpose of speeding up evolution; besides, organisms are not concerned in the least about the long term future. But the fact is that the lineages that followed this path ended up diversifying and multiplying themselves more than those that perpetuated the primitive system of asexual reproduction.

The solution that Konrad Lorenz proposes for explaining the changes in the level of organization is somewhat different. In effect, life as much as reflection represents qualitative leaps, which entail the appearance of new properties of matter, completely unknown on the planet up to that point. But those *flashes of brilliant light*, in the felicitous words of Konrad Lorenz, are not miraculous, since they are the integration of preexisting elements (Lorenz preferred the term *effulgence* to that of *evolution*, because originally *evolution* meant a foreseeable development, and *effulgence* is always something unexpected. In other words, life and consciousness do not simply represent the sum of elements prior to their appearance, but rather something more. Two plus two sometimes equals five, or five thousand. However, that *something more* is not to be taken as some kind of ethereal, intangible, mysterious force. Modern biology has renounced any form of animism or vitalism. Every new property, no matter how spectacular, emerges from the reorganization of elements available at the level of inferior organization: it is an *emerging property*. This scientific paradigm, which has allowed biology to dispense with *dark forces* in its explanations, is called *organicism*.

There is, therefore, a way to fit the emergence of life and the symbolic mind into the framework of modern science, from which, since the Baroque Revolution, all pretense of finality has been eliminated. But both events are so rare that they have occurred but once on Earth, and we were not present at the time to observe them. Without a doubt, they were highly singular events.

All that we can do at the present moment is to continue trying to see more clearly under what circumstances and from what foundations life and consciousness emerged; at present, we are very far from being able to produce an artificial life or artificial intelligence that can be truly considered analogous to the life and intelligence of the organic world. It is very doubtful that this will be achieved one day, but not on account of any metaphysical impossibility, but because research is not going in that direction. It is not, in the least, interesting to construct a robot that imitates the human being: we already have ourselves for that! What is worth our trouble is developing machines that simulate some useful properties (for humans) of life and human intelligence, even on a higher scale than natural creatures. That is to say, specialized machines, very efficient at what they do, and not artificial beings that know how to do a little of everything like organisms, which in reality are equipped by nature for living and reproducing, not for producing some serviceable function for man. It is indeed possible, and is being done, to build artifacts that make decisions and resolve problems for us, because they do it better and faster. But, of course, we humans are the ones setting the criteria for decisions.

Chapter X

Freemen or Slaves?

The Fourth Blow

Sigmund Freud was considered the author of the definitive blow that knocked man off the pedestal on which he thought he stood. The first blow was delivered by Copernicus when he claimed that Earth was not the center of the universe. The second was the discovery by Darwin that we descended from the *monkey*, and the third, the blow that Freud himself would have struck, was to delve into the recesses of consciousness in order to discover that down there, in the subconscious, is where our darkest desires (without our knowing it) are brewing, the ones that really impel us in life.

But another blow was yet to come, this time a low blow, and it was delivered in two consecutive years. In 1975, Edward O. Wilson published the book titled *Sociobiology: The New Synthesis*, and a year later, Richard Dawkins's book, *The Selfish Gene*, appeared. The essence of both premises is this: the relationships with our fellow creatures are regulated by the degree of parentage; the more consanguineous we are, the greater will be our disposition to help others. Wilson believes that on this very simple basis, an entire science of social behavior can be constructed: sociobiology. Bees and other hymenopterous social insects, like ants, and also termites (that are more distant in evolution), live in colonies (hives, anthills, termite societies) where only one individual is reproduced, which is the queen. The workers forgo having offspring of their own and instead sacrifice themselves (including death) for the queen's other daughters, their sisters, or to be precise, for the entire colony. When a bee stings us and dies because of it, it is behaving, to all appearances, in the most contradictory manner to its *interests* (and those of its genes) that one can imagine.

The explanation for such absurd behavior (as admirable as a kamikaze bee may seem to us) lies in this particular type of insect reproduction, where the females have two sets of genes (they are *diploids* like us, with one set of chromosomes from the father and another from the mother). On the other hand, the males only have one set (they are *haploids*). As a result of this singular system of genetic inheritance, a female worker bee shares more genes with her sister (three quarters) than with her own children (half) or her brother (one fourth). Therefore, instead of having children, the female workers *have* sisters.

In the case of these social insects (which are Wilson's area of specialization) it seems that everything balances out. In fact, it was the geneticist William D. Hamilton who was the author of this line of reasoning, but Wilson goes farther and maintains that this regulation of parentage would also be applicable to all other animals, and even in a possible human sociobiology. In this way, the dream of integrating the social sciences with the theory of evolution (the neo-Darwinian version) would be fulfilled.

But if what causes us to practice altruism with our relatives are the many genes we have in common, why not immediately dismiss individuals and only reason in terms of genes? Then, like Richard Dawkins, we can elaborate on the metaphor of the *selfish gene*: in reality, genes have *interests* and *intentions*. Therefore, they sometimes contravene the convenience of the individuals who carry them and obligate them to perform favors for other individuals, or even to give their lives for them. This apparent altruism is, in fact, reinforced and not really a free choice. It is *as if* genes manipulate our desires to their own advantage. It is *as if* they use us. It is *as if* we are mere vehicles for transporting them on their eternal journey through time. It is *as if* they make use of us in order to reproduce copies of themselves or give preference to their copies, which are carried by other bodies, including sacrificing our lives if necessary. What parents wouldn't give their lives for their children? Finally: why not discard the *as if* and immediately affirm that we are in the service of our genes? The low blow of sociobiology and of the metaphor of the selfish gene to human pride consists of assuming that it is the genes, and not the mind, that are in control. The ship's captain believes he is in command of the vessel because he is at the controls, but others sketch out his course: some infinitesimally small passengers called genes.

Edward O. Wilson and Richard Dawkins are pure neo-Darwinists, because they are convinced that natural selection is evolution's motor (singular in nature) and the mechanism responsible for the adaptations of living beings. For the neo-Darwinists, most of the characteristics of organisms are adaptations. Because behavior is also a form (and a very important one) of adaptation, behavior would have been an object of selection in animals too. Following the discovery of the existence of genes, it was learned that they determine in great measure the phenotype of individuals, including not only the physical appearance but also the behavior of the phenotype. But can the characteristics of human behavior be dealt with as if they were anatomical characteristics or physical traits? Must we speak of human behavior as a complete whole or can we break it down into units of behavior, each one of them determined (or codetermined) by a different gene? Are there specific genes for every one of our interests, inclinations, vocations, abilities, or sexual preferences? Faced with the propensity of sociobiologists to answer this last question in the affirmative, other scientists, like Richard C. Lewontin, Steven Rose, and Leon J. Kamin, shout in a loud voice: "NO!"

But for those of us who have a tin ear or little vocation for being a craftsman or artisan, we are convinced that it is something genetic and there is nothing to be done about it, no matter how diligent we are about attending the Music Conservatory or the School of Fine Arts. Many think there are people more gifted than others for learning languages, as is borne out daily in classes or for sports. That boundless penchant for do-it-yourself tasks, cooking, hunting, collecting, soccer, dance, women, men, etc., etc., etc., that we observe in our acquaintances (and in us), how much of it is innate? Is it true that little boys, by nature, like playing soldiers and little girls prefer playing with dolls? How much truth is there to it that men are better at handling maps and blueprints? If this is so, are we safer in a plane piloted by a man or woman? Are women generally better at verbal communication and personal contact, that is to say, in jobs dealing with the public?

Mammals' ova are much more valuable than spermatozoa for the simple reason that in a population, the proportion of one to the other is tremendously unbalanced. Moreover, a lengthy period of time has to pass before a female mammal carrying a fertile ovum (or several in species with multiple offspring) returns to a condition of fertility. That's why the sexual patterns of male and female mammals respond to different *interests* and are studied from the point of view of the difference in *value* between the ovum and spermatozoa. Is that a reasonable perspective for humans as well? Does the male *gravitate toward* fertilizing the greatest number of females possible, while the female, on the other hand, *aspires* to stability and security (independent of one another since both behave, more or less reluctantly, in accordance with the conventions of their culture)?

Among social mammals, there is frequently a very direct relationship between the legal status of the male and the size of his offspring; the higher up the male is in the group's hierarchy, the more probabilities he will have for having access to females with ova ready for fertilization. In many human cultures, this arrangement is also, or was also, provided for. Does that mean that the man, out of instinct, seeks power, and the woman seeks the one with power? If it is not true that the male is by nature more promiscuous than the female, how does one account for the clientele of prostitution and pornography being, universally, male? Is it simply a question of education and culture? These questions are not, of course, an effort to justify machismo, but to see if it has any biological basis, whether large or small; but in any case, and this must be very clear, biology will never be the goddess that decrees our range of values.

Each one has his own response to these and other questions or something like it, but many think that, at least, some personality traits are inherited. We know, too, that certain mental illnesses, unfortunately, are also hereditary to some degree, like schizophrenia. The question of real moral import is this: if it is true that there are specific genes behind many of our personality traits, and since advancement is being made with all due speed in knowledge of the genome, will we one day, perhaps not too far off, come to know which genes they are? If that is the case, will human beings be labeled, and some discriminated against, because of their genetic constitution? These are serious questions deserving of a little bit of our concentrated attention.

• • •

Social Darwinism

Those who are opposed to sociobiology fear seeing the ghost of social Darwinism resuscitated, that is, the idea that for the good of the species, class warfare or warfare among peoples is justified. Turning the argument around, the rich (individuals and countries) would be, by virtue of having imposed themselves on the weak, the strongest, the best adapted. If the traits of human behavior are so rigidly programmed, will the socially victorious be the best there is in intelligence, temperament, and morality? Social Darwinism is concerned with the *harmful* consequences that a slackening in natural selection would have for the species, allowing the least gifted to reproduce as much or more than the superior members of the human race. It goes without saying that helping the losers in the struggle for life means going against the act of natural selection, and therefore going *against nature*.

But social Darwinism was, in fact, a trap. It proclaimed the equality of opportunity and laissez faire as a political doctrine. It claimed that the fittest would emerge from the struggle among individuals, the leaders who must guide society for the good of all. But, in truth, social Darwinism did not want natural selection to result in equality of opportunity, for example, in the field of education. It was not a matter of all children having the same environment for developing their aptitudes and seeing who was best, but instead it was taken as a fact that this competition had already taken place long ago, and that the dominant classes and peoples had already sufficiently demonstrated their overwhelming superiority.

The presumed natural difference between some humans and others was not only on an intellectual scale, but also on a moral level. The students at Oxford and Cambridge were already, prior to entering the university, the best and brightest. A gentleman was not only better prepared than his servant or a *savage*, but he was also, by nature, more valiant, just, honorable, and decent. The life of the inferior classes and peoples was ruled more by their contemptible and chaotic passions, lacking in any moral limitations, than by their reason. Like intelligence, morality had also evolved, and not all (peoples, individuals, sexes) had achieved the same level. The product of the long and beneficial activity of natural selection could not be lost in the short-term with anti-natural egalitarianism, which could place the future of humanity in danger. If the least morally and intellectually evolved individuals came to power, what would become of us?

In his first edition of *The Origin of the Species*, in 1859, Charles Darwin defined natural selection as "the struggle for life." The definition of natural selection was later completed with the phrase "the survival of the fittest," coined in 1864 (in *Principles of Biology*) by the English philosopher and sociologist Herbert Spencer (1820–1903), and thus the words "the struggle for life or the survival of the fittest" stuck. It was this expression of Spencer's ("survival of the fittest") that allowed Darwinism to move to sociology, and therefore, social Spencerism must be spoken of rather than social Darwinism [certainly, Spencer defended the evolution of the species before Darwin, but his thought was Lamarckian; after *The Origin of the Species*, Spencer also accepted natural selection as (another) explanatory mechanism for evolution].

There was an important English philosopher and political theorist, long before Darwin's time, who also interpreted human societies as the result of the struggles among individuals. This is Thomas Hobbes (1588–1679), author of *Leviathan* (1651). Hobbes was impressed by Galileo's method (he met the Italian scientist in 1636 in Arcetri) for reducing a system to its elements and mechanically explaining it in terms of movements of bodies, and he wanted to do the same with human society, which he broke down into individuals so that he could then understand the whole; in other words, the social machine could be studied in the same way that physics studied the celestial system. Upon adopting a mathematical approach, Hobbes moved away from Bacon's method, which (as we saw) consisted of experimentation and induction without any previous model.

Hobbes's point of view is similar to the mechanistic reasoning that Galileo used for explaining natural systems, because for Hobbes, social manifestations are simply a consequence of the interactions among individuals; they do not lie in the nature of the individuals themselves, taken separately, but are the result of their relationships. Individuals do not have, in principle, an interest in living in society (and they lack moral concepts), but they see themselves forced to do this because of the limitation of resources, since all aspire to the same things, and there is not a sufficient amount available. This gives rise to aggression, insecurity, distrust, and hatred. Such passions turn the human being, in his natural state (*status naturae*) and forced by circumstances, into a wolf with respect to other humans (*homo homini lupus*), even though he may not be competitive by nature.

Thomas Hobbes's own father, an Anglican parish priest, was a violent man who beat another clergyman to death with his fists at the door of the church, and feeling obligated to leave, he abandoned his three children to the care of his brother. Thomas was born prematurely because of the shock his mother experienced on seeing enemy soldiers enter the town, and an important part of his life was spent in exile because of the English Revolution. Hobbes, apparently, was not on good terms with either of the two parties in the conflict: King Charles I and the Constitutionalists (the first lost his head in 1649). All these manifestations of human violence managed to influence his thinking.

Hobbes assumes the supremacy of the right (*jus*) of the individual, that is to say, his freedom, based on law (*lex*), or in other words, obligation; but to be able to live securely, men decide to live in society (*status civilis*), and to do that, they freely come to an agreement to transfer their rights to the absolute state, which then turns into a monster of unlimited powers comparable to Leviathan, a biblical beast. Hobbes's Leviathan is a superorganism, a system composed of all individuals grouped together. Leviathan (*commonwealth* in English, *civitas* in Latin, or state) is like an artificial man, even though of greater proportion and strength so that he can offer the functions of protection and defense for which he was created. The Crown would be like an artificial soul, which gives life and movement to the social body.

In nature, Hobbes's *bellum omnium contra omnes* (war of all against all) is inevitable, not because of the inclination on the part of individuals, but because of the scarcity of resources. This is how Darwin's natural selection is produced, and how the survival of the fittest, which Spencer defended, is guaranteed.

If in the origin of evolutionary theories, there is an influence from philosophers and sociologists like Hobbes and Malthus, social Darwinism or Spencerism seems to have paved the road in reverse and transferred the principles of biology to sociology. The German socialist philosopher Friedrich Engels was a Darwinist, but that did not stop him from observing the danger of this two-way reasoning, in which definitive individualistic political theories are converted into laws of obligatory application to human societies after their passage through the natural sciences.

Let's try to strip away all ideology from our scientific work in the search for the biological bases for human conduct. Let's begin by asking ourselves: do they exist?

• • •

The Brain's Integrated Circuits

For behavioral psychologists, with Skinner at the head, all that needs to be investigated in animals and humans are the events that modify the probability of responses. An individual's behavior would be, for this environmentalist school, merely the outcome of the strengthening or weakening of the voluntary responses that the subject has experienced during his life, stemming from the rewards and punishments he has received. We are only the product of our backgrounds, because we are born without any programmed behavior: it is life itself which programs us, and education based on rewards and punishments (or on the lack of reward).

The extreme opposite position (of which some radical sociobiologists are accused) would be to say that all relevant aspects of animal conduct, and *personality* in the case of *persons*, are conditioned to a greater or lesser degree by the genes. We would thus come into the world with highly rigid programmed designs of conduct; basically, we would be preprogrammed.

Ethologists stand somewhere in the middle of both extremes. For them, preprogramming exists in the nature in which animals know the world. The data of what is called extrasubjective reality (the world called real) reach them in a filtered form, so that the only things that are perceived are those things that hold an interest for the subject. Some motivating stimuli trigger immediate reactions in the individual, but internal instincts (drives) also exist that lead them to seek out the necessary stimuli so that specific behaviors are produced (especially when those behaviors have not been exercised for a long time). Moreover, genetic programs exist that permit learning and channel it. Not everything is assimilated, only what is appropriate for the individual's survival and reproduction. In some very striking cases, learning takes the form of typography, and the animal is forever marked by a kind of imprint that it receives at a susceptible moment in its life.

But how can we know if all of this preprogramming also exists in humans? Eibl-Eibesfeldt has studied human universality, that is to say, patterns that show up in all cultures, even in those so remote that they had never been *contacted* by Westerners. Some of these behaviors are also found in unweaned babies and deaf children, and those youngsters who are blind; these groups, unfortunately, are barely able to interact with others. What's more, some of these behaviors are also found in primates. These universal patterns are, probably, innate: one is born with them.

According to research conducted by human ethologists, preprogramming exists in the human being and is not very different from that which is found in those animals that are closer to us in characteristics. For starters, many of our expressions (body language) are innate, as Darwin himself had already observed. There is also a genetic component in human aggression, especially when a stranger enters the picture. Hierarchy and territory are, as in other species of social animals, a source of conflict.

But in no way does this mean that aggression is inevitable and must be accepted as fatalism, like a permanent and eternal curse. On the contrary, we are generally not a very aggressive species except, as has already been noted, when ideologies clash. In fact, humans carry very strong inhibitions against murder. Just the way we have a repertoire of aggressive gestures at our disposal, we also have in our species, as in other social species, gestures of salutation and appeasement that forestall the aggressor (but which, naturally, cannot work when killing takes place from long range, where one does not see the victim).

Of course, there is nothing better for putting an end to aggression than education for coexistence based on mutual understanding; in this way, the stranger (whether a foreigner or someone who is different) ceases to be a cause for fear or anger, and the "stranger equals enemy" mind-set is broken. As the geneticist J. B.

S. Haldane put it, if we can't avoid favoring our brother and mistrusting the foreigner, the best thing to do is to preach universal fraternity, the way religions do. If we are convinced that we are all brothers, it will be easier for us to get along together. There is a very recent experiment that feeds our hopes of realizing that goal.

The researchers who carried out the experiment are Robert Kurzban, John Tooby, and Leda Cosmides, from the Center for Evolutionary Psychology, at the University of California, in Santa Barbara. Evolutionary psychology is a branch of psychology that applies the logic of the theory of evolution as a tool for tackling problems. The researchers at the University of California were interested in the basic categories in which humans classify, automatically and imperatively, their fellow beings when they encounter each other for the first time. Traditionally, it was believed that the three *primitive* or *primary* dimensions that we use for categorizing people unfamiliar to us were sex, age, and race. These three dimensions would be independent of one another and sufficient, meaning that our minds would compute them without any need for taking in any additional information; for example, they would suffice for recording in our minds that the new neighbor is a young black woman.

The conditions under which race would not be a factor in categorizing (codifying) individuals has been the object of a lengthy, but unsuccessful, search. If it were an irrefutable fact that race is a natural category for our brains, there would be reason for concern, because the simple act of classifying individuals into two groups (us and them) predisposes human beings to favor their own group over the other. Even so, we would have the moral obligation to combat racism with all our might, even our neurocomputational engine. But are we certain this is really so difficult?

The aforementioned evolutionary psychologists asked themselves why race is a primary dimension of the human being. The reason for our minds being preprogrammed to establish these three categories is that they are good indicators for predicting behavior in others, and therefore, they have adaptive value in the social environment in which the human being has to defend himself/herself. In other words, we know, more or less, what we can expect from a child, adult, or elderly individual, but what predictive value is there in knowing race? Above all, keeping in mind that there is so little variation in our species that one cannot properly speak of races, since most genetic variation arises *within* populations and not *among* them. And in addition to that, one must consider that those slight changes that exist in the human likeness follow one another geographically in very moderate gradations (pigmentation, for example, does not abruptly pass from the white Scandinavian to the sub-Saharan black, but rather there are intermediate skin tones in the lands situated between one region and another). Finally, since the mobility of prehistoric populations in all likelihood was unavoidably limited, what probability would a person then have had of meeting up with another person markedly different in appearance?

As a consequence, these authors put forth the hypothesis that no part of human cognitive architecture was adapted for codifying race, and that the observations that previously seemed to indicate that race was a category that was automatically encoded, really meant something else. Prehistoric hunters and food gatherers lived in bands, and the bands would enter into conflicts with one another fairly frequently (something that has been observed in modern human populations with this type of economy and also occurs among chimpanzees). Under such circumstances, the ability to detect alliances probably had a great adaptive value, and thus a neurocognitive mechanism would have emerged through evolution that figures out which alliance an outsider belongs to by predicting his behavior. Consequently, it is the group that would be codified by category and not race. The results obtained from classic experiments pointing to race as a primary dimension sprang from the fact that race coincided with the group in the experimental designs, something that, no doubt, probably also happened in some prehistoric frontier regions where two different human populations met (like the Eskimos and Indians today, for example), but which was probably a rare event.

If this hypothesis were correct, it should have been able to experimentally prove that membership in a group carries more weight when it comes to cataloging strangers than the differences in the color of their skin or other racial characteristics.

Kurzban and his colleagues carried out a series of experiments aimed at proving his hypothesis, which basically consisted of having some observers witness a discussion between two groups of people. There were whites and blacks on both teams, and association with one team or the other was made explicit by several types of unequivocal signals. Next, the observers were questioned about the discussion with questions like: who said what? If the observers categorized the participants by race, the mistakes in attributing a statement to the wrong person would be produced among individuals of the same race, but the exact opposite was demonstrated. Confusion arose more frequently among persons of the same team.

The final result of the experiment was that it had taken less than four minutes for racial barriers to be erased and group identities to take root. Skin color and other traits that differentiate populations from one another do not establish natural categories when it comes to classifying human beings, just as the color of one's hair or the shape of one's nose among Spaniards does not. The brain, on the other hand, really takes quick notice of who (whether a brunette, redhead, or blond) goes with who in the game of life. Without having to resort to finely tuned experiments, we believe that sports teams demonstrate that rivalry between races only exists when the teams are racially homogeneous, but on the other hand, no one (fortunately) notices the skin *color* of the player who scores a goal if he is from their team and wears their *colors*.

As we see, there are arguments that point to the existence of certain preprogramming in categorizing strangers, but important preprogramming, nevertheless, seems to exist in human learning, especially in the area of language. According to the linguist Noam Chomsky, all languages have a common basis, because universal laws of language exist. Children learn with neuronal structures in their brains that allow them to learn something as difficult as a language at an early age, long before they know how to multiply, for example. One would say that language is something natural to them, and on the other hand, math is not. But this programming is open-ended and does not encode which language the child will absolutely speak; it only determines the basic structure, which is, according to Chomsky, common to all languages.

Long before all of this, Darwin thought in terms not very different from those of Chomsky's: "Horne Took, one of the founders of the distinguished science of philology, notes that language is an art, the same as distilling beer or baking bread; it would have been better to compare it to writing; but certainly it is not an instinct, because language has to be learned. It greatly distinguishes itself, nevertheless, from all the ordinary arts because man demonstrates an instinctive tendency for speech, as we see in the babbling of our children, who, on the other hand, show no instinctive tendency to distill beer or bake bread, or to write."

The theory developed by Chomsky in the 1950s received quick

support from the neurologist Eric Lenneberg, who observed that there seemed to be a noticeable hereditary component in some children with language deficiencies, because certain cases are repeated in particular families. The importance of these studies lay in their suggestion of the existence of something like a *language organ*, in other words, neuronal networks that are already formed before the child begins to speak (precisely to make it possible for him to learn a language). This preprogramming was probably innate or, put another way, most likely belonged to the *hardware* and not the *software* of the brain. The opposite hypothesis is that language ability is not an independent faculty, but instead forms part of a broader group of skills that we could designate as *general intelligence*, and which can be measured by psychological testing like the well-known *intelligence quotient* (IQ).

Some findings have recently been obtained that could prove Chomsky right. In 1990, a large family came to the attention of researchers; in this particular case, through three generations, a special kind of difficulty with language presented itself, when it came time to producing it as well as understanding it. The affected parties seemed unable to apply any elementary rules of language that a normal child automatically produces; it was more than a speech problem. In all, twenty-seven individuals were genetically studied (out of a total of thirty-seven), of which fifteen displayed linguistic disorders.

The difficulty with language for the family in question could not be the consequence of low intelligence either, because some of the affected members achieved scores on intelligence tests that were within the norm and were superior to the scores of other members of the same family who were not affected by the long-standing problem.

In 1998, the researchers linked the defect in that family's linguistic problem to a small segment of chromosome 7 (an *autosomal* or non-sexual chromosome), which they named SPCH1. Finally, in October 2001, a team headed by Cecilia S. L. Lai and Simon E. Fisher made public the discovery of the gene responsible for the linguistic problem; the identification was made possible thanks to a member of another family having the same problem, which presented a chromosomic alteration that split that gene (a *translocation*). This is gene FOXP2, and the disorder is caused by a mutation that affects a base (adenine instead of guanine) in the chain of the gene's nucleotides (and in a single strand of the DNA double helix). The result is the change in an element in the chain of amino acids of the protein, which the FOXP2 gene codifies, possibly altering its function.

The gene seems to emphatically exhibit itself in the fetal development of the brain (among other places), and it is clearly known that the homologous gene in mice exhibits itself in the development of the cerebral cortex in embryos. The conclusion drawn by the researchers is that the effects of the mutation take place at a crucial moment during embryogenesis, producing the abnormal development of neuronal structures that are important for language and speech.

There are, therefore, reasons for suspecting that the mutation does not simply cause *general* cerebral disorders that *also* affect language, but that it, in fact, impedes the complete formation of Chomsky's so-called *language organ*.

The Human Genome Project, therefore, is beginning to bear fruit in the area of *cognitive genetics*. We will soon see how far it can advance along this route in learning more about the biological bases of human language.

How To Fry a Couple of Eggs

When, in a previous chapter, we took up the matter of self-replication or autoreplication, we made reference to the principal problem that it presents, which is the question of an internal blueprint. An organism cannot construct another just like itself if it does not have a plan from which to work, where the final design is sketched down to the last detail. Instead of a sketch, however, the internal blueprint could be printed out in text format, requiring, of course, much more paper, knowing as we do that a picture is worth a thousand words. This is true for the possible autoreplicating machines of the future, but in the case of living organisms with sexual reproduction, whether plants or animals, the problem is quite different. It is not now a question of an individual having an internal blueprint, a basis on which it could make a copy of itself, but rather that of facing another situation: a male sexual cell and a female sexual cell link up to form an egg cell or zygote that contains, in its nucleus, all the information needed for constructing an adult; in other words, the new individual will assemble itself from its first cell in a process that is called development.

How is the miracle of development possible? It was explained by an old theory of *preformation*, which stated that the zygote or egg cell of an animal held the complete contents for the formation of the adult with all its characteristics, but in a microscopic version: literally, the zygote would house a miniature of the adult with all its parts. The process of development would be, in fact, a simple expansion of the zygote, a mere change in scale (the preformationists termed this form of development *evolution*). Confronting the doctrine of preformation was that of *epigenesis*, which held that the embryo is not preformed in the egg, but is configured (from a shapeless mass) through a series of stages in which the adult aspect gradually takes root. That orderly development is produced by a force called *vis essentialis*, which certainly has to be distinct for every species. Modern developmental biology has demonstrated that both schools were partly correct and also partly mistaken. In effect, there is no *homunculus* (a miniature version of man) in our zygote, but rather, we are formed through embryogenesis; in this instance, the epigenesists were right. But they were wrong in that there is no external force or *vis* that acts upon or channels embryogenesis to produce an adult human, monkey, or frog; instead, the orderly development is attributable to a preformed element, the DNA genetic blueprint.

The Italian scientist Lazzaro Spallanzani studied in great detail the world of cells, and as we have already seen, was a precursor of Louis Pasteur in his opposition to the theory of spontaneous generation. Spallanzani was interested in understanding the role that spermatozoa, discovered in 1677 by the Dutch scientist Antoine van Leeuwenhoek, had in reproduction. Along with his friend, the Swiss naturalist and philosopher, Charles Bonnet (1720–1793), Spallanzani expanded on his own version of the theory of preformation. According to his interpretation, the origins of all living organisms, past, present, and future, were created by God and encased in the first female of every species. The new individuals, which are born, are developed from an egg that exists from the beginning of time (even though the father's semen, on contact, provides a stimulus for development). In other words, every female contains within her sexual cells an infinite series of preformed individuals, which guarantees the inalterability of the species in the future.

Bonnet and Spallanzani thus found a way around the obstacle of infinite regression, which was so difficult for us when we posed the question of how self-replicating automatons could construct themselves. In Bonnet and Spallanzani's terms, the automaton would have to contain an egg with another miniature-sized automaton inside. That miniature automaton would increase in size until it

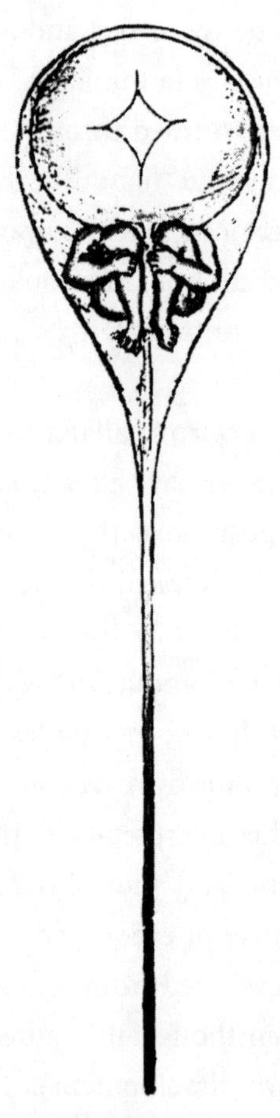

Figure 24: At the end of the seventeenth century, a preformationist illustration of the human body contained in a spermatozoon.

became an *adult* automaton; but then, how could there be a third generation of automatons? Here's the solution: the automaton offspring also contains an egg with an automaton, and this in turn, contains another, etc. In other words, like Russian nesting dolls, the first automaton has to carry the eggs (encased in each other) of all the automatons that are going to exist (as many as there are generations foreseen until the end of the series of automatons).

In the case of living organisms, the Creator would have already seen to it that our egg was inside our mother's egg, which in turn was inside our grandmother's egg, which was inside our great grandmother's egg, and as far back as Eve.

We now know that the zygote contains the father's genes as well as the mother's, but at the root of the idea that personality is inherited in all its details, lies the preformationist model, which says that an adult in miniature is found inside the first cell. Not, of course, with an adult form that can be recognized looking through a microscope, but indeed in the form of plans, which in the end describe in every detail the design of the building or the machine that one wishes to build; a blueprint is a projection of something, just as the zygote is the blueprint of an adult. The development, therefore, would be a process of reading and executing those plans with the maximum fidelity possible. Every gene would correspond to a trait, in the same way that every detail of the blueprint is translated into a part of the building or machine. Or put another way: everything that is in the result is in the blueprint, and everything that is in the blueprint becomes reality.

The great Russian geneticist Theodosius Dobzhansky explained the same erroneous concept another way. When it is asserted that we inherit from our father or mother a particular personality trait or physical characteristic, we are imagining that biological inheritance is similar to property inheritance by way of man-made laws, which regulate the transfer of assets from parents to children. In short, since a new being is forged in the same zygote, all the inherited traits that will configure the adult are already present (*preformed*) and unalterable. Upon inheriting a gene from our father or mother, we will be inheriting, at the same time, the trait encoded in the gene. If that were so, our character traits and talents would be totally determined from the cradle (from the zygote, in fact), and we would be forever locked in to them. But, Dobzhansky added, the laws of biological inheritance are not like the laws that govern material inheritance. We do not get our characteristics in the same way that a chair or painting is bequeathed to us from our grandfather's estate.

A human ovum has a diameter of 0.1 of a millimeter and weighs 0.0015 milligrams. The factor by which we have to multiply the weight of an ovum to obtain that of a normal adult person is astounding: about 50 billion. H. J. Muller, cited by Dobzhansky, calculated that all the ova that led to the entire world population (which was around 2.5 billion people at the time) would easily fit inside a 1-gallon container (less than 4 liters). The corresponding number of spermatozoa would take up a much smaller space, smaller than that of an aspirin! In the zygote, the spermatozoa and ova, despite their great difference in size, carry the same number of genes. The sum of all the genes of all the sexual cells that led to the entire world population (2.5 billion people during Muller's time) would fit inside a vitamin capsule! However, that tiny mass is humanity's genetic patrimony, the inheritance of some 3.8 billion years of evolution, and the foundation for the future of our species.

To increase weight by a factor of 50 billion, the zygote has to construct a body with resources acquired from the environment, including the necessary energy for growth and life. There is no operative agent present to read a blueprint contained in the genome to fabricate a body, but genes have to function like a blueprint for development, that is to say, a set of instructions for growth. Therefore, the genome cannot be an autodescription of the type of blueprint, but instead has to be something like a recipe, that is to say, a set of instructions that explains how to use elements in the environment to achieve a result that is predictable only up to a certain point.

The recipe for two fried eggs with potatoes is not the same thing as a photo of a plate of fried eggs with potatoes, but rather a very streamlined description of a sequence of steps (which we assume have no need of explanation . . .). The success of the final *dish* depends on accurately following the steps as outlined, in their correct sequence, and using the proper ingredients, even though there are no two dishes of fried eggs exactly alike; the result varies according to the type of eggs, potatoes, oil, amount of salt, etc. that are used (plus many other purely random factors that also affect the result). Even though the *genetic recipe* channels development and makes it possible, it does not figure in the infinite details with respect to the materials needed for going from the fertilized ovum to the adult. That is why, not even twins that come from the same ovum and the same sperm cell (and are, therefore, clones) are physically equal. Quality of environment influences the phenotype that is constructed from the genotype. By *quality of environment,* we mean not just the material component, but the spiritual as well. In other words, twins are not psychological clones either.

Don José's Historic Reason

Spain's best philosopher, José Ortega y Gasset (1883–1955), wrote profound treatises on human character. For him, life is given to us not because of anything we have done, but because it is there, and one day we must deal with it. But neither is life given to us as a fait accompli, an already finished product, but as something that each one of us has to make for himself/herself; in other words, life takes work. We have to be continually active (life is conjugated in the present participle) and realize that we are obligated to constant decision-making. What happens in life is that man realizes that any decision-making becomes impossible if he does not possess some conviction about the reality of things, what they are, what other men are, and what he himself is. Only from those convictions can he favor one action over another, and since to live is to decide, man—thanks to his convictions—can live.

During his time, Ortega thought that a long cycle of faith in reason that had begun at the end of the sixteenth century had come to a close and that, in turn, had negated the religious faith that had characterized the Middle Ages, a faith founded on the belief that revelation was sufficient in order to understand the world. The Baroque faith in reason knew no limits, and the solution to all problems was considered within its reach. The birth of the Modern Age in the fifteenth and sixteenth centuries was not produced without labor pains, and the twentieth century would witness, according to Maestro Ortega, a similar crisis.

What could have been the cause of this uneasiness that Ortega contemplated? It was the failure of the sciences, natural and social alike, in their attempt to understand human nature. The basis of the Baroque faith in reason lies in the fact that Western man believes that the world has a rational order that coincides with the structure of the human intellect, and it can be mathematically expressed. The scientific method, adopted by physics, chemistry, and later on, biology, was rooted in that fundamental belief, which is the same one that the so-called spiritual sciences eventually adopted as well. But Ortega believes, as happened before with religious thought, that neither physical reason nor biological reason has anything to say about man, and he therefore finds it imperative to construct a vital or historic reason. For Ortega, the logic for why none of the sciences succeeded in understanding human nature is quite simple: rather than having a nature, man has a history. Our behaviors are not conditioned by our genes but by the historic confluence of time in which the convictions we now hold and those held by others before us have gradually merged.

Ortega was writing at a time of dreadful suffering in the world, and the inability of science to put a stop to it must have tormented him like so many other thinkers of the terrible twentieth century (without, so it seems, things changing very much in the newly born twenty-first century). If faith in science was excessive, that is not to say that it has not come far, very far, in its knowledge of human nature, which we believe plays an important role in biology. Ortega's radical environmentalism seems excessive today, but the maestro was right when he said that ideas condition the way we perceive the world, the way we regard one another and the rest of mankind.

For many years, Konrad Lorenz studied how the cognitive apparatus, with which we are born, works, a tool that is formed by the sensory organs and central nervous system; that biological system is responsible for our internal imaging of the world, which of course is not a facsimile of the real world. But, moreover, Lorenz recognized the existence of another filter through which we see reality, another cognitive apparatus that was probably formed by the culture that we acquire when we socialize. Culture, a product of history (as Ortega pointed out), also makes us see certain things while blinding us to other things; it is also a filter for stimuli (superimposed on the biological filter).

That is the reality we have to confront: according to Ortega, man always has *to be rooted* in some belief; beliefs constitute the underpinnings of our lives. We are accustomed to saying that *we have* definite ideas, "but as for our beliefs, more than having them, we are them."

Richard Dawkins not only formulated the theory of the selfish gene, which concedes the leading role in an individual's existence to a few *replicates* (or *replicators*), genes, which nest in the interior of their cells. For us humans, Dawkins, conceived of another metaphor no less unsettling, which refers to a few replicators that reside in our brains. It is what he called *memes* (a word that is a hybrid of *imitator* and *gene*). The term "meme" is understood as a reference to any cultural element susceptible to being imitated. Examples of memes given by Dawkins are "tunes or sounds, ideas, slogans, fashion, methods of constructing vessels or building arches."

Dawkins's daring idea is that memes duplicate themselves like genes and, thus, skip, just as genes do, from one body to another, perpetuating themselves. Memes are, if one wishes, guests of our brains. When someone plants a meme in another brain, he literally *infects* it, converting that brain into a vehicle of propagation for the meme. It is something very similar to what happens with viruses that parasitize cells, and which cannot be reproduced without them, or to give a more contemporary example, memes behave like computer viruses. If organisms, according to Dawkins, are mere "machines for the survival of genes," we humans are also (and this is what makes us different) *machines for the survival of memes*.

Memes can travel from one brain to another in many ways, beginning with conversation and observation; and of course, modern means of communication (written, audiovisual, digital) have broadened their capacity and speed of growth by leaps and bounds. They can also remain dormant in many kinds of media, like books, an inscription on a stone monument, videotapes, photographs, CDs, and computers; but to replicate themselves, they definitely need to have access to a human mind. In fact, the length of time a meme survives does not depend so much on its longevity, that is to say, on what it has inscribed onto one particular medium, be it paper, stone, computer, or a brain, but rather on its productiveness, or in other words, its efficiency at the time of self-replication. The more copies it produces, the easier it will be for it to continue existing and reproducing itself in more and more brains.

But what is really troublesome is not that memes introduce themselves and are replicated in the human brain, but that they can have *interests* different from the interests of the people they colonize. The analogy to the selfish gene is clear in this instance, because in the same way that genes can make use of bodies for their own survival and proliferation, memes can also be harmful for the persons *infected* and, despite that, propagate themselves (behaving like real parasites).

There have been ideas that have led many humans to sacrifice themselves, for different causes (as we are witnessing these days in suicide attacks), or dying rather than renouncing them (as in the case of the first Christians), or simply, by not procreating. In all these cases, genes (and people) are harmed by memes, even though those destinies (death or celibacy) are freely chosen.

What makes a meme reproduce itself is its expression or phrasing, the affect that it produces, as much as or more than the concept it contains; speaking in biological terms, what counts is its *phenotype,* in the same way that a gene is a unit of information that materializes in the phenotype. Genes are also selected on the basis of their expression and not for themselves, since obviously natural selection cannot *see* the sequence of nucleotides that constitute the gene's syntax. We already saw how the meme, *remove plants at night from the bedroom,* continued to multiply itself an infinite number of times since its creation, in this case, because the expression of the concept that plants breathe oxygen at night is coated with a component of fear that makes the idea quite pervasive.

The problem with memes is that some are beneficial while others are harmful, and others are of noneffect. If it is true that plants represent a danger to children, the meme *remove plants at night from the bedroom* is beneficial, as is also the one about vaccinating children against tuberculosis. But harmful memes multiply themselves despite their injurious nature because their phenotype holds an attraction, or because they are associated with other memes that have a good *image*; the alcohol and tobacco industries make sure that their products have a psychological association with the good life (the very thing that suffers most because of their harmful effects).

There is, nevertheless, an important difference between genes and memes: the fact that, in the end, the memes that find their way into our heads are those that we find attractive, and which we let pass. They proliferate because we like them, not like genes, which come to us through heredity at birth, and against which we are helpless (still, but not for long). Is this the salvation of our free will?

Richard Dawkins ends his book *The Selfish Gene* with a song that pays homage to human freedom. His reasoning goes like this: selfish genes do no long-range planning, but rather act in accordance with their short-term *interests*, and the shorter, the better. But what benefits them at any given moment can be their ruin in the end. Memes, the other replicators, generally act the same way and are also unable to foresee the future, because they are very simple replicators, unconscious and blind. Dawkins hopes that another unique quality of humans will be the possibility of being able to offer a true altruism, not only unmistakable, but totally unselfish. But, in the worst case scenario, we humans are at least able, thanks to our mental abilities, to be selfish and visionary at the same time, and to understand that it is sometimes better to forego the immediate benefit for the sake, at a later date, of obtaining the higher reward (or of avoiding the pernicious effects of a momentary possible good). We can even associate and cooperate with other *systems of survival* for sharing costs and benefits. Thanks to our ability to *simulate* the future in our heads, we are the only beings capable of disobeying the replicators (be they genes or memes).

We are, as a consequence, the only free beings. But when Dawkins speaks of *us* being free, Daniel Dennett asks: who is the *us*? The very memes that reside in our brains and shaped our minds, which *are* our minds. These reflections, which threaten to drive us mad, are the same ones that led Ortega y Gasset years ago to assert that while we possess ideas, *we reside* in our beliefs.

It is at this precise moment in time that we are turning our desperate gaze to the biological bases of human behavior, premises much maligned by those authors who come down on the side of culture as the critical factor. Can we not trust evolution, which has made us human and endowed us with the ability to separate what is beneficial from what is harmful? We believe the answer is yes.

It is good that research on the genetics of human behavior (cognitive genetics) is ongoing. We will do a better job of dealing with the disorders that confront us in living together once we understand them better. This research will take years, and the results are yet to be seen. In our opinion, there is no need to fear what we will encounter. We think that in the end, we will learn that genes program us to be free. That means that our genes do not rigidly map out the way we live life, and if we educate our children to be self-reliant, tolerant, and nonviolent, they will be able to learn to live in peace and prosperity in a more just . . . a more humane world.

Chapter XI

Gaia

The British chemist James E. Lovelock and the American biologist Lynn Margulis perceived that a narrow relationship existed between life and the physical framework in which it evolved, and they developed a hypothesis called Gaia (after the Greek goddess of the earth), which achieved wide popularity, almost taking on the appearance of a *religion for atheists*. The essential idea of the Gaian hypothesis is that life did not evolve *on* Earth, but rather evolved *with* Earth; put another way, over the course of time, life and Earth coevolved, mutually influencing each other. According to this model, the life that dwells on our planet would combine with air, soil, and water to form a singular and very complex system, which can be seen as a single organism that has the capacity for making our planet a suitable place for life.

The Gaia hypothesis owes much to a Russian scientist little known to the general public, a geologist from Saint Petersburg named Vladimir Ivanovich Vernadsky. He was the one who coined the term *biosphere* and, for the first time, conceived of the totality of life as a unity that forms a fine layer on the exterior of the planet Earth, even though it is condensed into individual, physically separate beings, which was what had been observed up to that time: one could not see the forest for the trees. That mass of life, which surrounds the planet, benefits from the energy and nutrients that it finds in its environment.

The biosphere is interconnected to the other spheres, of course: the lithosphere, hydrosphere, atmosphere. Therefore, the biosphere is really an ecosphere, a broader concept that includes the totality of the planet's life (the *biota*), plus the other spheres with which organisms obviously exchange materials, and that is the principal merit of the Gaia hypothesis: to make it clear that Earth, sea, air, and life have a common history and destiny. The word ecosphere also carries another very important meaning: it is a system, a group of interconnected elements.

Systems have emerging properties (the adjective itself bears unsettling resonances). Emerging properties not only depend on the elements that comprise the systems, but on how those elements are related to one another. Biological systems embody each other like the Russian nesting dolls: cell, tissue, organ, system (again), organism (up to this point, we have not even finished examining the field of pure biology), and lastly, ecosphere (the largest of all the planet's living systems, which also concerns geology); we'll skip the ecosystem, because it only represents one area of the ecosphere.

We know that the systems that lie beneath the individual level are parts of a whole, that is to say, they are in turn elements of a superior organized system. Why, then, should we not assume that individuals are parts of the same type of system as an organism (though on a higher level)? What will we name that system that integrates all living things and, at the same time, the physical medium that surrounds them? Superorganism. What shall we call that superorganism? Lovelock and Margulis proposed calling it Gaia.

Up to this point, we are all in agreement. We can even go farther without losing the consensus. Let's say that Gaia has a history: it is not the product of the circumstances that are found at one specific moment of life, but rather, it is something that belongs to the passage of time. In order to belong, it has to change, or adapt, because circumstances change. Going one step farther, we reach the very edge of the abyss: let's say that life has not limited itself to adapting to planetary changes, but that life itself is the fountain of change on occasion; to put it another way, life is a geological agent. We now have a version of Gaia that is, in general, well received and does not spark any polemics. On the contrary, all of us who aspire to an ever increasingly better communication among the life and Earth sciences are grateful to Lovelock and Margulis for their idea.

Now we come to the abyss. Like a good organism, does Gaia have *homeostasis*, that is to say, is it self-regulating? Confronted with a change in the medium, does it react with a *feedback* mechanism? If something is missing in the medium (for life, of course), does Gaia come up with the goods (positive *feedback*)? If there is an excess of something, does Gaia reduce it (negative *feedback*)? Moreover, does Gaia improve the conditions for life over the course of time? Whoever answers these questions in the affirmative can consider himself aligned with the most radical or *toughest* version of Gaia, that of its creators.

Those homeostases are produced in such temporally long cycles that it is not easy to compare them. Practically the same thing happens with the notion of progress in the complexity of the biosphere. Paleontologists are still discussing whether there has been a basic tendency over time (but with ups and downs) in the increase of quantity and diversity of marine life since, shall we say, the Cambrian period, up until today.

There is, however, an aspect of Gaia that is even more troubling: that is its possible *teleological* character. Does Gaia have a *telos*, a purpose, an objective of its own? How do you feel when you think that it is only one element, among many, of a larger organism, something like being one cell in your own body, and that superorganism called Gaia has been going its own way for a very long time and will continue doing so? In your body, the tissue cells die and are replaced (except in the central nervous system, generally speaking) by others without your being aware of it, because your body is constantly reconstituting itself in order to stay alive. If it could think, a cell of your skin would consider you immortal compared to its own very short life. Does that kind of pantheism, the feeling that you form part of something larger and longer lasting (Gaia), console you over the fact that you are condemned to die just as the cells of your body are, one after another?

In principle, Gaia is not a conscious entity (even though we will discuss this later), but that does not mean that it can't exist and have its own interests. All organisms have a *telos*: to stay alive (and to transmit life), although admitting that does not assume any form of *vitalism*, nor that plants, animals, or bacteria have to be conscious because of it. The important concept is not that of teleology, but rather the one offered by teleonomy, a variant of the term that has been used by authors like Jacques Monod, Konrad Lorenz, and many others in referring to the structures of organisms that fulfill functions, which are useful. Those structures perform services, the reason for which they are adapted to their ends. Just like our artificial tools (beginning with the stone ax), teleonomic structures are objects equipped with a design, plan, and *intention*. Darwin's great revolution was to show how teleonomic structures could surface in the course of evolution without any need for anyone to plan them, thanks to the mechanism of natural selection. As Daniel Dennett has recently written, what Darwin proposed was an algorithm that in some mechanical way, and without any aim per se, carries out very complicated functions of calculus (all the *engineering* and *biotechnology* of animals). Teleonomical structures are those that distinguish a bee from a mountain. The topographical relief is not an adaptation; Earth's peaks and valleys do not fulfill functions, and they have no design or plan. Teleological (*intentional*) thinking is not applied outside of biology. The moon produces tides, but it doesn't occur to anyone that this is its function; the moon simply exists and tides are its effects. But if Gaia is a superorganism, where are those adaptive, or teleonomical structures, which characterize all the individuals that we recognize as such?

It is not easy to identify the existence of something that is the equivalent of adaptations in the biosphere, because although adaptations have functions, those functions are subordinate to the general good of the organism. The teleonomical structure that is the eye cannot be independently conceived, could not survive on its own; in fact, we don't even see with our eyes: we do it with the brain (a person who has a significant lesion in the primary visual area of the cerebral cortex, on the occipital lobe, is blind, even though her eye may have suffered no damage at all). In other words, the *telos* of the adaptation (its objective, its end, its purpose, its sense, its function) is subordinate to the organism's general *telos* (which is perpetuation). Can it be said that there are parts of the biosphere whose existence can only be understood in terms of *service* to the well-being of the group? In a system as complex as the biosphere, some parts cannot exist without the others, that much is clear, but can we correctly say that plants exist *in order to* make life possible for other organisms that do not have the capacity, as we do, of photosynthesis? Does Gaia not assume the return to a vision of an intelligently ordered world like the one Saint Thomas offers in his fifth way? Is Gaia a scientific version of the idea of natural harmony?

On the other hand, if Gaia is a superorganism equipped with a *telos*, how was it created historically? We could say that organisms collaborate with each other in the system's homeostasis to maintain constancy in the conditions of life, and even to improve them; we could even imagine that they evolved to more effectively cooperate, but how could organisms cooperate if the reason for their evolution is precisely because they compete among each other in *the struggle for existence,* according to Darwin's declaration?

The altruism that is recognized in the species in benefiting Gaia merits a separate comment, because it is indeed true that behaviors called *altruistic* or cooperative are observed among individuals of particular species. At one time, it was said that such altruism was realized for the *good of the species*, a notion that has been, as we saw, substituted by one that says for the good of the genes (in a series of theories that begins with J. B. S. Haldane and continues with William D. Hamilton, John Maynard Smith, Richard Dawkins, and Edward O. Wilson). If the *altruistic gene* has been replaced by the *selfish gene*, how are we to believe that all the individuals in the biosphere are paddling in the same direction?

With considerations like these, the scientific idea of a superorganism like Gaia will continue to *wane* and, little by little, descend (or ascend?) into the realm of a metaphor. Since we think in terms of spheres, why not lend an ear to an even more daring metaphor? The French Jesuit priest and paleontologist Pierre Teilhard de Chardin invented the concept of the *noosphere*, the planet's thinking envelope. A sphere that certainly has become reality with the *Internet*: billions of interconnected brains, consciously working together, which are slowly converging toward a kind of omniscient universal mind.

Chardin, the *discoverer* of the noosphere, and Vernadsky, that of the biosphere, consulted and mutually influenced one another, according to what Chardin himself says in a letter: "Vernadsky was in Paris during that period and I saw him frequently." Gaudefroy was the geological father who encouraged Teilhard to read Vernadsky's *Geochemistry,* because the Russian also aspired to make a single science of all the phenomena of life and Earth. It was around 1925 when Chardin coined the term and concept of noosphere, after the model of Vernadsky's biosphere (Earth's living envelope), and used it for the first time in *L'hominisation* (*Hominization*) (Paris, May 6, 1924).

So it was those two visionary scientists who dreamed about developing a total synthesis and were in frequent contact in Paris around 1925. Vernadsky was sixty-two at the time, and Chardin forty-four. Gaia was not a product of their association, but now, three quarters of a century later, it is possible for us to think that with the two spheres brought together (*superimposed*), we would have a biosphere integrated with a noosphere, an authentic and conscious Gaia. To a certain extent, that could also be considered the principal conclusion of Chardin's work: How intimately do his ideas resemble the mysticism that beats in the heart of the Gaian hypothesis, the light that shines in the pupil of the Gaian believers!

Gaia is a beautiful metaphor, but it is not needed. In the final paragraph of *The Origin of the Species,* Charles Darwin tries to have us understand that the scientific idea of evolution is more wonderful than any myth or metaphor regarding our origins: "There is grandeur in this view of life, with its several powers, having been originally breathed into a few forms or into one; and that whilst this planet has gone cycling on according to the fixed law of gravity, from so simple a beginning endless forms most beautiful and most wonderful have been, and are being, evolved." That planet that gave birth to us was held sacred by the Greeks as the goddess Gaia, which is the *Terra Mater* of the Romans; the Basques call it *Amalur,* Mother Earth.

Bibliography

Alexopoulos, Constantine J., and Mins, Charles W., *Introducción a la Micologia,* Omega Barcelona, 1985

Allègre, Claude J., and Scheneider, Stephen H., "La evolucion de la Tierra," in: *Investigación y Ciencia* (December1994), pp. 36-45.

Anguita, Francisco, and Arsuaga, Juan Luis, "¿Es Gaia una teoría adelantada a su tiempo o una broma vitalista?" in: *Documentos del XI Simposia sobre la Enseñanza de la Geología (XI S.E.G.),* Santander, 2000, pp.20-26.

Arsuaga, Juan Luis, *El collar del neandertal. En busca de los primeros pensadores,* Temas de Hoy, Madrid, 1999.—, *El enigma de la Esfinge. Las causas, el curso y el propósito de la evolución, Areté,* Barcelona, 2001.

Arsuaga, Juan Luis, and Martínez, Ignacio, *La especie elegida. La larga marcha de la evolución humana,* Temas de Hoy, Madrid, 1998.

Atkins, Peter, and Jones, Loretta, *Química. Moléculas, Materia, Cambio,* Omega, Barcelona, 1998.

Ayala, Francisco J., Rzhetsky, Andrey, and Ayala, Francisco, "Origin of the metazoan phyla: Molecular clocks confirm paleontological stimates," in: *Proceedings of the National Academy of Sciences of U.S.A.,* 95 (1998), pp.606-611.

Bazzaz, Fakhri A., and Fajer, Eric D., "La vida de las plantas en un mundo enriquecido en CO2," in: *Investigación y Ciencia* (March 1992), pp. 6-13.

Bonner, William A., and Castro, Albert J., *Química orgánica básica,* Alhambra, Madrid, 1976.

Chen, Jun-Yuan, Huang, Di-Ying, and Li, Chia-Wei, "An early Cambrian craniate-like Chordate," in *Nature,* 402 (1999), pp. 518-522.

Childress, James J., Felback, Horsts, and Somero, George N., "Simbiosis en las profundidades marinas," in: *Investigación y Ciencia* (July 1987), pp. 78-85.

Cohen, Richard, *History of Life* (2nd Edition), Blackwell Scientific Publications, Londres, 1990.

Cuenot, Claude, Pierre *Teilhard de Chardin. Las grandes etapas de su evolución,* Taurus, Madrid, 1967.

Darwin, Charles, *El origin de las especies por medio de la selección natural,* Calape, Madrid, 1921.

—, *Diario del viaje de un naturalista aldrededor del mundo, Calpe,* Madrid, 1921.

—, El origen del Hombre y la selección en relación al sexo, Ediciones Ibéricas, Madrid, 1966.

—, *Autobiografia y cartas escogidas,* Alianza Editorial, Madrid, 1997.

Dawkins, Richard, *El gen egoísta,* Salvat, Barcelona, 1985.

Deacon, J. W., *Introducción a la Micología moderna,* Limusa, México D.F., 1988.

Delius, Christoph, Gatzemeier, Matthias, Sertcan, Deniz, and Wünscher, Kathleen,

Historia de la Filosofía. Desde de la Antigüedad hasta nuestros días, Könemann, Colonia, 2000.

Dennett, Daniel C., *La peligrosa idea de Darwin*, Galaxia Gutemberg-Círculo de Lectores, Barcelona, 1999.

—, *Tipos de mentes,* Debate, Madrid, 2000.

Dobzhansky, Theodosius, *The Biological Basis of Human Freedom,* Columbia University Press, New York, 1956.

Dubos, Rene J., *Louis Pasteur: free lance of science,* Little, Brown and Company, Boston, 1950.

Eibl-Eibesfeldt, Irenäus, *Etología. Introducción al estudio comparado del comportamiento*, Barcelona, Omega, 1974.

—, *El hombre prepogramado*, Madrid, Alianza Editorial, 1977.

Eldredge, Niles, Reinventing Darwin. *The great debate at the high table of evolutionary theory*, Wiley, New York, 1995.

Fernández Álvarez, Manuel, Historia. *La Universidad de Salamanca*, Universidad de Salamanca, Salamanca, 1991, pp. 9-44.

Flammarion, Camille de, *L'atmosphère. Météroloqie populaire*, París, 1888.

—, *La atmósfera. Los grandes fenómenos de la naturaleza*, Montaner and Simón. Barcelona, 1902

García-Alcalde, Jenaro L., "Evolución biótica y geográfica en el Paleozoico inferior y medio," *Registros fósiles e Historia de la Tierra*, Editorial Complutense, Madrid, 1997, pp.119-142.

Gillispie, Charles Coulston (editor), *Dictionary of Scientific Biography*, Charles Scribner's sons, New York, 1980.

González, Daniel H., Iglesias, Alberto A., Podestá, Florencia E., and Andreo, Carlos S., "Metabolismo fotosintético del carbono en plantas superiores," in: Investigación y *Ciencia* (April 1989), pp. 76-83.

Gould, Stephen Jay, *Desde Darwin: reflexiones sobre historia natural*, Herman Blume, Madrid, 1983

—, *El pulgar del panda*, Hermann Blume, Madrid, 1983.

—, *La vida maravillosa, Crítica*, Barcelona, 1991.

Govindee and Coleman, William J., "Cómo producen oxígeno las plantas," in: *Investigación y Ciencia* (April 1990), pp. 50-57.

Hidalgo, F. Joge, *La nueva Materia*, Servicio de publicaciones de la Universidad de Extremadura, Cáceres, 1991.

Horgan, John, "En el principio. . . " in: *Investigación y Ciencia* (April 1991), pp. 80-90.

Huxley, Julian, and Kettlewel, H. D. B., Darwin, Salvat, Barcelona, 1994.

Jacob, François, *El juego de lo posible*, Grijalbo Mondadori, Barcelona, 1982.

Janvier, Philippe, "Catching the first fish," in: *Nature*, 402 (1999), pp. 21-22.

Kardong, Kenneth V., *Vertebrados. Anatomía comparada, función, evolución*, McGraw-Hill, Madrid, 1999.

Knoll, Andrew H., "El final del Eón proterozoica," in: *Investigación y Ciencia* (December 1991), pp. 26-47.

Kurzban, Robert; Tooby, John and Cosmides, Leda, "Can race be erased? Coalitional computation and social categorization," in: *Proceedings of the National Academy of Sciences of U.S.A.* , 98 (2001), pp. 15.387-15.392.

Lai, Cecilia S. L., Fisher, Simon E., Hurst, Jane A., Vargha-Khadem, Faraneh, and Monaco, Anthony, P., "A forkhead-domain gene is mutated in a severe speech and language disorder," in: *Nature*, 413 (2001), pp. 519-522.

Lars, Borg, Connelly, James N., Nyquist, Larry E., Chi-Y., Shih, Weismann, Henry, and Young, Reese. "The age of the carbonates in martian meteorite ALH84001," in Science, 286 (2001), pp. 90-94.

Laskar, Jacques, "La Luna y el origen del hombre," in: *Investigación y Ciencia* (July 1994), pp. 70-77.

Lehninger, Albert L., *Bioquímica*, Omega, Barcelona, 1981.

Levington, Jeffrey S., "La edad de oro de la evolución animal," *Investigación y Ciencia* (January 1993), pp 44-53.

Lewontin, Richard C., Rose, Steven, and Kamin, Leon J., *No está en los genes. Crítica del racismo biológico*, Grijalbo Mondadori, Barcelona, 1996.

Lezcan, Antonio and Alexander I., "Oparin: apuntes para una biografía intelectal," *Orígenes de la vida. En el centenario de Aleksandr Ivanovich Oparin*, editorial Complutense, Madrid, 1995, pp. 15-39.

Lopez Piñero, José María, *Ciencia y técnica en la sociedad española de los siglos XVI y XVII*, Labor, Barcelona, 1979.

—, "La ciencia en la España del Barroco," *Arte y Saber, La cultura en tiempos de Felipe III y Felipe IV*, Ministerio de Educación y Cultura, Valladolid, 1999, pp. 149-176.

Lopez, Piñero, José María, Navarro Brotons, V., and Portela Marco E., *Materiales para la historia de las ciencias en España: S. XVI-XVII*, Pre-Textos, Valencia, 1976.

Lorenz, Konrad, *Sobre la agresión: el pretendido mal*, Siglo XXI, Bilbao, 1972.

—, La otra cara del espejo, Círculo de Lectores, Barcelona, 1990.

Macphail, Euan M. *The Evolution of Consciousness*, Oxford University Press, Oxford, 1998.

Madigan, Michael T., Marlinko, Jonh M,. and Parker, Jack., *Brock Biología de los Microorganismos* (8th edition), Prentice Hall, Madrid, 1998.

Margaleff, Ramón, *Ecología* (50th edition), Planeta, Barcelona, 1992.

Mayr, Ernst, *Así es la Biología*, Debate, Madrid, 1998.

Meglitsch, Paul A., *Zoología de invertebrados*, Hermann Blume, Madrid, 1978.

Miller, Stanley, L., "The First Laboratory Synthesis of Organic Compounds under Primitive Earth Conditions," *The Heritage of Copernicus: Theories A Pleasing To The Mind*, The M.I.T. Press, Cambridge, Massachusetts, 1974, pp. 228-242.

Monod, Jacques, *El azar y la necesidad*, Orbis, Barcelona, 1985.

Montero, Francisco, "Evolución prebiótica. El origen de la vida," *Origen y Evolución: desde el Big Bang a las sociedades complejas*, Fundación Marcelino Botín, 1999, pp. 199-255.

Montero, Francisco, and others, *Evolución prebiótica: el camino hacia la vida*, Eudema (Ediciones de la Universidad Complutense de Madrid), Madrid, 1993.

Oparin, Alexander I., *El origen de la vida*, Editores Mexicanos Unidos, México D.F., 1976.

Orgel, Leslie E., "Origen de la vida sobre la Tierra," in: *Investigación y Ciencia* (December 1994), pp.46-55.

Ortega and Gasset, José, *Historia como sistema*, Espasa-Calpe, Madrid, 1971.

Pasteur, Louis, "Acerca de las generaciones espontáneas" (transcipción de la conferencia pronunciada el 7 de abril de 1864 en La Sorbona, traducida por Miguel de Asúa), in: *Ciencia Hoy*, 59 (October/November, 2000).

Peretó, Juli G., *Orígenes de la evolución biológica*, Eudema (Ediciones de la Universidad Complutense de Madrid), Madrid, 1994.

Pinker, Steven, "Talk of genetics and viceversa" in: *Nature* 413 (2001), pp.465-466.

Prigogine, Ilya, and Stengers, Isabelle, *La nueva alianza. Metamorfosis de la ciencia*, Alianza Universidad, Madrid, 1983.

Ron, Eugenia, and Sobota, Tomás (editors), *Manual de Botánica*, CD-Rom, Universidad Complutense de Madrid, Madrid, 2001.

Romer, Alfred S., *Man and the Vertebrates*, Penguin, Harmondswoth, 1954.

—, *Anatomía comparada. Vertebrados*, Interamericana, México, 1973.

Rosnay, Joël de, Orígenes de la vida, Ediciones Martínez Roca, Barcelona, 1970.

Sagan, Carl., "La búsqueda de vida extraterrestre," in: *Investigación y Ciencia* (December 1994), pp.62-69.

Sceilaher, Adolf, Bose, P.K., Pflüger, F., "Triploblastic animals more than 1 billion years ago: trace fossil evidences from India," in: Science 282 (1998), pp. 80-83.

Shapin, Steven, *La revolución científica. Una interpretación alternativa*, Paidós, Barcelona, 2000.

Shen, Yanan, and others, "Isotopic evidence for microbial sulphate reduction in the early Archaean era," in: *Nature*, 410 (2001), pp.77-81.

Shu, Degan G., Zhang, X.-L., and Conway Morris, Simon, "Lower cambrian vertebrates from South China," in: *Nature* 402 (1999), pp.42-46.

Sipper, Moshe, and Reggia, James A., "La autorreplicación de las máquinas," in: *Investigación y Ciencia*, October, pp. 14-23.

Taylor, G. Jeffrey, "El legado del proyecto Apolo," in: *Investigación y Ciencia* (September 1994), pp. 12-19.

Vega, Lope de, *Rimas humanas y otros versos (Edición y estudio preliminary de Antonio Carreño), Crítica, Barcelona, 1998.*

Wang, D. Y. C., Kumar, S., and Hedges, S. Blair, "Divergence time stimates for the early history of animal phyla and the origin of plants, animal and fungi," in: *Proceedings of the Royal Society of London. Part B Biological Sciences* 22 (1999), pp. 163-171.

Wendt, Hebert, *Tras las huellas de Adan: la novella del origin del hombre*, Noguer, Barcelona, 1973.

White, Tim D., "Prehistoria del canibalismo," in: *Investigación y Ciencia* (October 2001), pp. 50-57.

Wilson, Edward O., *Sociobiología: La nueva síntesis*, Omega, Barcelona, 1980.

—, *Consilience, La unidad del conocimiento*, Galaxia Gutemberg/Círculo de Lectores, Barcelona, 1999.

Wilson, Jean D. (editor), *Harrison. Principios de medicina interna*, Interamericana-McGraw-Hill, Madrid, 1992.

Wray, Gregory, A., Levinton, Jeffrey, S., and Shapiro, Leo H., "Molecular evidence for deep precambrian divergences among Metazoan *phyla*" in: Science 274 (1996), pp. 568-573.

Yochelson, Ellis L., "From farmer-laborer to famous leader; Charles D. Walcott (1850-1927)", in: *GSA Today* 6 (1996), pp.8-9.

INDEX

C

O

P